LIQUID SCINTILLATION COUNTING

LIQUID SCINTILLATION COUNTING

Volume 1

Proceedings of a Symposium on
Liquid Scintillation Counting,
University of Salford
September 21–22 1970

Editor: A. Dyer (University of Salford)

HEYDEN & SON LTD
London · New York · Rheine

Heyden & Son Ltd., Spectrum House, Alderton Crescent, London NW4 3XX.
Heyden & Son Inc., 225 Park Avenue, New York, N.Y. 10017, U.S.A.
Heyden & Son GmbH, Steinfurter Str. 45, 4440 Rheine/Westf., Germany.

© Heyden & Son Ltd., 1971.

All Rights Reserved. No part of this publication may be reproduced, stored in a retrieval system, or transmitted, in any form or by any means, electronic, mechanical, photocopying, recording, or otherwise, without the prior permission of Heyden & Son Ltd.

Library of Congress Catalog Card No. 70-156826

ISBN 0 85501 053 3

Made & printed in Great Britain by
Stephen Austin and Sons Limited, Hertford

Contents

Preface		vii
Chapter 1	Chemiluminescence as a Problem and an Analytical Tool in Liquid Scintillation Counting D. A. Kalbhen	1
Chapter 2	The Absolute Method of Measurement of Carbon-14 Activity by 4π Liquid Scintillation Counter T. Radoszewski	15
Chapter 3	Emission Spectra of Liquid Organic Scintillators E. Langenscheidt	23
Chapter 4	Scintillations in Liquid Helium P. B. Dunscombe	37
Chapter 5	A Liquid Helium Polarimeter of Unique Design J. Birchall, C. O. Blyth, P. B. Dunscombe, M. J. Kenny, J. S. C. McKee and B. L. Reece	43
Chapter 6	The 4π Liquid Scintillation Method for Activity Measurements of Electron Captive Nuclides A. Tada and T. Radoszewski	49
Chapter 7	Optimisation Techniques for Computer-Aided Quench Correction M. I. Krichevsky and C. J. MacLean	55
Chapter 8	The Processing of Liquid Scintillation Spectrometer Data Using a Desk-Top Computing System M. A. Williams and G. H. Cope	69
Chapter 9	A Comparison of Computer-Input Methods Used to Process Liquid Scintillation Counting Data D. S. Glass and T. L. Woods	79

Chapter 10	A New Gelifying Agent in Liquid Scintillation Counting *A. Benakis*	97
Chapter 11	Measurement of Radiation Effects on Thyroid Cell DNA Synthesis using Tritiated Thymidine *W. R. Greig*	105
Chapter 12	Quantitative Studies of Enzymes and Drug Actions in Cells and Microslices using ^{14}C-labelled Substrates *G. A. Buckley, C. E. Heading and J. Heaton*	123
Chapter 13	Critical Remarks about Current Trends in Liquid Scintillation Counting *D. A. Kalbhen*	127
Index		131

Preface

The technique of liquid scintillation counting has undergone extraordinary advances since its first conception some twenty years ago. Perhaps the most striking aspect of this advancement is the breadth of utility the technique enjoys. Physicists, whose interests lie in the fundamentals of the energy transfer process inherent to the technique or in the calibration of radioactive standards, share a certain common ground with medical researchers using isotopes in a variety of biochemical studies.

Advances in instrumentation have to keep pace with the complexities of sample preparation and the handling of large accumulations of data and this has generated a need for computerisation. So, in keeping with contemporary science and technology, statisticians and computer scientists are becoming involved in liquid scintillation counting.

This book is a record of an International Symposium on Liquid Scintillation Counting held at the University of Salford on September 21st—22nd, 1970. The Symposium was intended to bring together people of multidisciplinary interests in an environment where discussion could be promoted and mutual problems analysed and, perhaps, solved.

In retrospect, I am grateful to the support the Meeting received, and in particular to the contributors and participants from Europe for whom the Meeting was intended.

I would like to acknowledge the assistance of the Nuclear Chicago Division of G. D. Searle Ltd. and Koch-Light Laboratories Ltd., which greatly contributed to the social side of the Meeting.

Mr. B. R. Lumb of Nuclear Chicago was the most willing and capable of co-organisers and deserves my grateful thanks.

It is a pleasure also to acknowledge the excellence of the Chairmanship of the individual sessions as carried out by Dr. J. B. Birks (University of Manchester), Dr. G. G. J. Boswell (University of Salford) and Dr. B. W. Fox (Paterson Labs., Christie Hospital) which did so much to help the success of the Symposium.

Finally, I would like to express my thanks to my secretary, Mrs. M. Shulkind and to Messrs. D. Brown, G. G. Hayes and R. P. Townsend, who did so much to ensure the smooth running of the Meeting.

A. D.

Chapter 1

Chemiluminescence as a Problem and an Analytical Tool in Liquid Scintillation Counting

D. A. Kalbhen

Institute of Pharmacology, University of Bonn, Germany

INTRODUCTION

Within the last 10 years the liquid scintillation counting technique has become the generally preferred method for the measurement of low energy β-emitting radioisotopes such as tritium, carbon-14, calcium-45 or sulphur-35. Due to its high sensitivity and efficiency and to the great progress which has been made in standardising the instrumentation and data handling equipment this technique has increased significantly in applicability and popularity.

Nevertheless there may occur certain reactions and factors which can disturb radioactive measurements and one of these factors is chemiluminescence.

In the past, chemiluminescence in liquid scintillation counting has been called the 'bête noir'.[1] Indeed it caused a great deal of trouble and has contributed to many errors in quantitative determination of radioactive material by the liquid scintillation technique. Only in recent years the cause and the physico-chemical nature of these luminescence phenomena, as well as the factors involved, have been thoroughly investigated.[2-6] So today it is quite possible to calm down or to tame this 'bête noir' or to avoid any contact. In this respect the main part of this chapter will deal with the problems of chemiluminescence which may occur in liquid scintillation counting, and we shall report some hints and notes on how to overcome these problems in tracer experiments.

On the other hand chemiluminescence reactions are of interest and importance in organic and bio-organic chemistry and can be used for analytical purposes in non-radioactive assays with great benefit. For these assays liquid scintillation counters are quite adequate measuring devices permitting a highly sensitive quantification. Therefore a second and smaller part of this chapter is concerned with chemiluminescence as an analytical tool. I am sure that this note is of special interest for all researchers using liquid scintillation counters, as well as for the constructors and manufacturers of these instruments, because the analytical use of chemiluminescence reactions may extend to a broader application of liquid scintillation counters.

PART I

The basic concept of the liquid scintillation technique makes it necessary that the radioactive sample material is in *intimate* contact or in *actual solution* together with the phosphor solutes, in order to obtain maximum counting efficiency by maximum energy transfer and photon yield. The following solvents and solutes have most successfully been used in liquid scintillation counting:

SOLVENTS

TOLUENE, DIOXANE,
METHANOL, ETHANOL,
METHYL-, ETHYL-,
BUTYL–GLYCOL,
XYLENE,
TRITON X–100,
POLYGLYCOLETHERS.

SOLUTES

PPO, POPOP,
BBOT, BIS–MSB,
PBD, DIMETHYL–POPOP,
BUTYL–PBD,
NAPHTHALENE,

From among the many liquid scintillation solutions which are used in various laboratories all over the world we have selected the following 17 cocktails (Fig. 1.) with qualitative and quantitative varying composition and we have investigated their sensitivity and reactivity for chemiluminescence reactions.

It can be seen from this list that many scintillation solutions contain dioxane in order to make them miscible with water containing samples.

Unfortunately the range of sample materials, especially of biological samples, that can be dissolved in such solvents is severely limited.

Besides other methods of sample preparation some basic solubilisers, as listed below, are used to digest or dissolve such biological materials as, amino acids, polysaccharides, nucleic acids, proteins, cells and entire tissues.

BASIC SOLUBILISERS

AQUEOUS OR ALCOHOLIC KOH OR NaOH
HYAMINE-10 X, NCS, PRIMENE 81-R,
PHENYLETHYLAMINE, SOLUENE-100

Although initially the quaternary ammonium bases were regarded as 'universal' solubilising agents, there are some limitations and disadvantages such as severe quenching, colour formation or chemiluminescence, which may occur by the use of HYAMINE, NCS or SOLUENE.

For many years investigators have observed spurious counts and high background levels in liquid scintillation counting, especially when radioactive biological materials solubilised by strong organic or inorganic bases were measured.

This 'photo-luminescent effect' was first attributed to proteins reacting with the quaternary ammonium bases.[7,8] In recent years the causes of the unwanted chemiluminescence reaction have been somewhat better understood and investigated and will be reported here. It should be mentioned that phosphorescence or static discharges are not involved here and are outside of the scope of this work.

(1) PPO   4 g
    POPOP   50 mg
    Toluene   1000 ml

(2) PPO   2.6 g
    POPOP   80 mg
    Toluene   300 ml
    Methylglycol 500 ml

    [Butler, *Anal. Chem.* **33**, 409 (1961)]

(3) PPO   4 g
    POPOP   75 mg
    Naphthalene 120 g
    Dioxane   1000 ml

(4) BBOT   4 g
    Naphthalene 80 g
    Toluene   600 ml
    Methylglycol 400 ml

(5) PBD   133 mg
    Naphthalene 33 g
    Dioxane   1000 ml

(6) PPO   5 g
    bis-MSB   0.5 g
    Toluene   1000 ml

    [Permablend (Packard)]

(7) PPO   5 g
    bis-MSB   0.5 g
    Naphthalene 120 g
    Dioxane   1000 ml

    [Permablend (Packard)]

(8) Butyl-PBD   7 g
    Toluene   1000 ml

(9) Butyl-PBD   7 g
    Naphthalene 100 g
    Dioxane   1000 ml

(10) PPO   7 g
     POPOP   50 mg
     Naphthalene 50 g
     Dioxane   1000 ml

(11) PPO   4 g
     POPOP   0.2 g
     Naphthalene 60 g
     Ethylglycol 20 ml
     Methanol   100 ml
     Dioxane   1000 ml

     [Bray, *Anal. Biochem.* **1**, 279 (1960)]

(12) PPO   6 g
     Methylglycol 600 ml
     Toluene   1000 ml

     [Mahin and Lofberg, *Anal. Biochem.* **16**, 500 (1966)]

(13) NE 233
     (Toluene-based)

     [Nuclear Enterprises Ltd.]

(14) NE 250
     (Dioxane-based)

     [Nuclear Enterprises Ltd.]

(15) Toluene-based
     Scintillator solution
     for Soluene™

     [Prototype from Packard]

(16) Insta-Gel

     [Packard]

(17) PPO   5 g
     bis-MSB   0.5 g
     Toluene + Triton X–100
     (2 + 1)   1000 ml

Fig. 1

METHODS AND RESULTS

In our experimental procedure for quantitative determination of the intensity of chemiluminescence reactions 1.0 ml of the basic solubilising agent was added to 10.0 ml of scintillation cocktail or solvent in the counting vial and the light impulses were measured 30 s later over a period of 6 s in a liquid scintillation counter, at a temperature of 10°C. If not otherwise mentioned the high voltage of the photomultipliers of the instrument were adjusted in the same manner as for carbon-14 integral counting.

The chemiluminescence measurements were performed with Nuclear Chicago liquid scintillation counters (three channel models 725 and Mark II) and two Packard instruments (Tricarb 3003 and Tricarb AAA 3380/544).

As bleaching agents hydrogen peroxide (30%) and benzoyl peroxide (saturated solution in toluene) were used. All reagents were of 'analytical grade'.

Whilst investigating 17 scintillation solutions it was found that dioxane containing cocktails gave highest chemiluminescence values while toluene or toluene-methylglycol based cocktails had minor chemiluminescence intensities.

An example of the duration and decay of chemiluminescence, shown in Fig. 2, indicates that the count rates are initially very high, fall exponentially over several orders of magnitude within the first 30 mins but remain elevated for several hours over the background baseline.

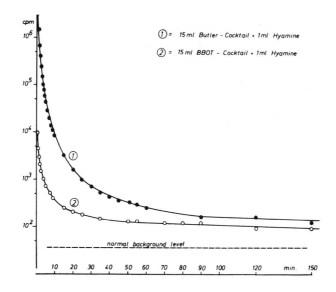

Fig. 2: Decay and duration of chemiluminescence reaction at 10°C in two scintillation mixtures containing 1.0 ml HYAMINE 10–X (counts on log-scale).

Because some liquid scintillation counters are operated with a refrigerated system while others run at ambient temperature, it was of interest to know the influence of temperature on the degree of chemiluminescence. For the investigation of the temperature dependence, 10.0 ml of our Butler Cocktail (No. 3) and 0.5 ml HYAMINE were mixed and measured at different temperatures. The results are shown in Fig. 3 and demonstrate that the chemiluminescence reaction is more intensive at higher temperatures.

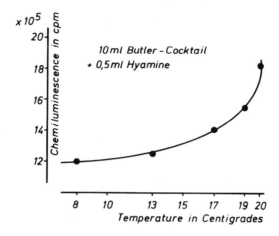

Fig. 3: Influence of temperature on the intensity of the chemiluminescence reaction.

It was further found that the different solubilising agents exerted quite different chemiluminescence intensities as may be seen from Table 1.

Table 1. Chemiluminescence of 0.5 ml Base and 10.0 ml Butler Cocktail.

HYAMINE	8.7×10^5 cpm
NCS	42.4×10^5 cpm
SOLUENE	61.2×10^5 cpm

This effect may be related to their different solubilising potency. While soft tissues and cells are easily digested by these solubilisers, harder materials such as cartilage, bone and collagen fibres will not dissolve completely. SOLUENE is the most potent of these solubilisers and in contrast to HYAMINE did not produce any colour formation.

As was demonstrated in our experiments the chemiluminescence reaction was always more pronounced when the solvents contained scintillator solutes. To investigate the effect and participation of solutes on the chemiluminescence reaction the photon yield of a mixture of dioxane and HYAMINE with increasing amounts of naphthalene dissolved in dioxane was measured. The results of this experiment are shown in Fig. 4.

Although a mixture of naphthalene and HYAMINE alone did not emit any light, there was a significant increase in chemiluminescence depending on the amount of naphthalene dissolved in the dioxane. These findings indicate that naphthalene itself may not react with HYAMINE, but it makes the dioxane more transparent, so that the photomultiplier of the counter will 'see' more light impulses. On the other hand it cannot be excluded that naphthalene (and this may be true for other scintillator solutes too) may participate in some, as yet unknown way in the chemiluminescence reaction. Naphthalene as well as the scintillator solutes may improve the energy transfer efficiency of the solvents.[9]

Investigations as to the cause and physico-chemical nature of chemiluminescence in liquid scintillation counting have made clear that organic peroxides react in an alkali

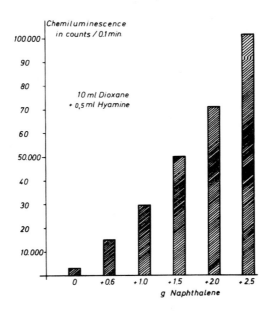

Fig. 4: Effect of naphthalene (dissolved in dioxane) on the intensity of chemiluminescence registered by a liquid scintillation counter.

medium to produce emission of light (see Hercules[5] and Gundermann[10]). Neither proteins nor other substances from biological samples appear to be involved in this phenomenon. To demonstrate that chemiluminescence is due to peroxides present in dioxane and toluene as contaminants, these solvents were shaken with small amounts of hydrogen peroxide and measured after addition of HYAMINE.[2,3] The result was an intense stimulation of light emission. As reported by Winkelman and Slater,[11] the increase of chemiluminescence was even more pronounced, when benzoyl peroxide rather than hydrogen peroxide was used.

During sample preparation for liquid scintillation counting, solutions of digested tissues, blood or urine in HYAMINE or NCS very often become coloured depending on the haem or pigment concentration in the digested material. This colour increased with the temperature and duration of the solubilising process. To avoid colour quenching, bleaching agents, such as hydrogen peroxide,[8] benzoyl peroxide[12,13] chlorine water,[14] sodium borohydride[15] have been used for decolourisation. Benzoyl peroxide was found to be the most satisfactory agent and the least quenching among those examined.[12]

Studies by Winkelman and Slater[11] and those in our laboratory have shown that mixtures of benzoyl peroxide, basic solubilising agents and scintillation cocktails will produce very intense chemiluminescence (Fig. 5). This can last not only for hours but for days and even weeks.

With regard to the fact that benzoyl peroxide may cause severe and long lasting chemiluminescence in scintillation mixtures containing *strong basic agents,* the use of this bleaching agent can be quite problematic.

Although there are quantitative differences in the intensity of chemiluminescence between basic solubilisers, such as HYAMINE, NCS, sodium and potassium hydroxide, our experimental results have indicated that an alkaline medium is essential for the

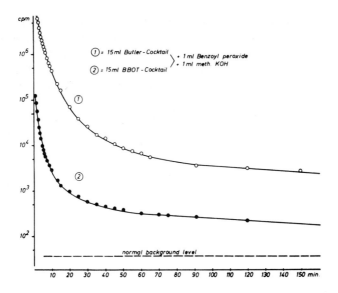

Fig. 5: Chemiluminescence decay curves of liquid scintillation mixtures containing benzoyl peroxide and methanolic KOH (counts on log-scale).

chemiluminescence reaction in standard scintillation solutions. Addition of acid to a neutral pH or lower than 7.0 will generally stop luminescence.

Chemiluminescence reactions in liquid scintillation mixtures are of quite low energy and may for this reason interfere even more in tritium counting. To demonstrate this a mixture of 1.0 ml benzoyl peroxide, 1.0 ml NCS and 15.0 ml solution D (Butler

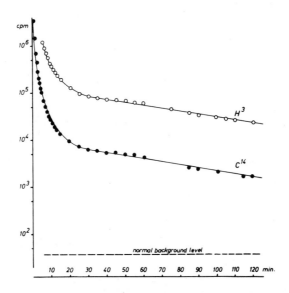

Fig. 6: Chemiluminescence curves of a mixture of 1.0 ml benzoyl peroxide, 1.0 ml NCS, and 15.0 ml Butler Cocktail measured at carbon-14 and tritium settings of the counter (counts on log-scale).

Cocktail) was measured at a high voltage setting of the counter for carbon-14 and tritium. As can be seen from Fig. 6, the chemiluminescence is much higher when measured in the tritium range, although the decay curves are quite similar in shape.

Until recently, chemiluminescence was known only to occur in an alkaline medium but, as we will see later, it is possible to reduce or eliminate chemiluminescence by just neutralising or acidifying the alkaline digest.

A few months ago during a comparative investigation[16] on sample preparation methods, solutes and solvents for liquid scintillation counting, a chemiluminescence was

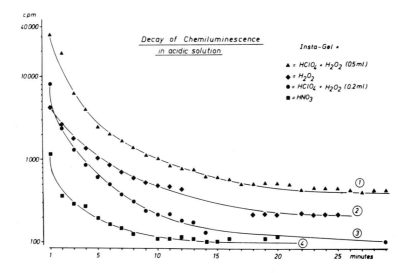

Fig. 7: Chemiluminescence from oxidising acids in Insta-Gel.

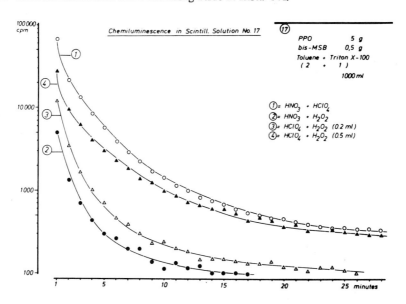

Fig. 8: Chemiluminescence from oxidising acids in emulsion cocktail No. 17 (see Fig. 1).

observed to occur when some biological material after wet combustion with perchloric acid and hydrogen peroxide were mixed with the emulsion cocktail Insta-Gel or the scintillation solution of Patterson and Greene.[17]

As may be seen from Fig. 7, further studies confirmed that the liberation of molecular oxygen from oxidising acids (such as nitric or perchloric acid) or from hydrogen peroxide in the presence of surface active gelifying agents (polyglycol ethers) can result in long-lasting chemiluminescence.

Quite similar effects were found in the toluene-Triton X-100 based cocktail of Patterson and Greene, as shown in Fig. 8.

CONCLUSION

To eliminate and overcome problems of chemiluminescence it is at first necessary to differentiate between chemiluminescence reactions in homogeneous, toluene or dioxane based cocktails and those in emulsion cocktails. For homogeneous scintillation solutions we recommend the following points:

1. In contrast to dioxane which is notorious for the formation of peroxides on contact with air, toluene-based scintillation mixtures generally contain much less peroxides and should be used preferentially in connection with basic solubilisers such as NaOH, KOH, HYAMINE, NCS.

2. The use of peroxides as bleaching agents should be avoided in basic solubilisers.

3. If biological samples are used, which had been digested in basic agents the samples should be neutralised to pH values equal or lower than 7.0 by the addition of acid. It should be mentioned however, that acid may increase quenching and may not always be sufficient to eliminate chemiluminescence.[11]

4. All counting samples of a pH higher than 7.0 should be checked for chemiluminescence effects in order to avoid counting errors. If chemiluminescence is found in the prepared counting sample it is advisable to store the sample at room, or higher temperatures until the luminescence has decayed to a tolerable level.

5. If dioxane-containing scintillation solutions have to be applied, it is advisable to use acidic solubilising agents or the perchloric acid/hydrogen peroxide technique of Mahin and Lofberg[18] for the digestion of biological materials.

When using emulsion cocktails in combination with the wet combustion technique of Mahin and Lofberg[18] or with other oxygen liberating acids we recommend a pre-alkalised (i.e. Insta-Gel containing 1 N sodium hydroxide) scintillation solution which brings the pH of the final mixture to a value of 5 to 6. This can be done without remarkably decreasing the counting efficiency.

Houtman[19] recommended BHT (di-*t*-butyl-4-hydroxy-toluene) as an antioxidant to reduce or eliminate chemiluminescence. Our results however, shown in Fig. 9, indicate that it is quite impossible to eliminate the luminescence completely by the addition of BHT even at concentrations up to 1 g per 10 ml scintillation solution; but as can be seen, BHT even contributes to chemical quenching, we therefore do not support the idea of Houtman.[19]

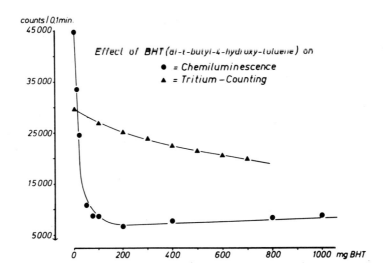

Fig. 9: Influence of BHT on luminescence and tritium counting.

PART II

Chemiluminescence as an analytical tool

Since the development of highly sensitive photodetectors and photomultipliers about 20 years ago, chemiluminescence phenomena in organic chemistry have been extensively studied. They have been found to occur in a great number of organic and bio-organic reactions. Although the reaction mechanism of many light emitting chemical processes is not yet completely understood, the various factors that may determine the quantity and efficiency of the light emission have been investigated and are relatively well known.[10]

Quantitative microchemical assay

The intensity of the luminescence reaction, for example, which occurs in a mixture of luminol and hydrogen peroxide, is known to be dependent on the concentration of metal salts that catalyse the reaction. Based on this and other chemical reactions, many procedures have been developed which can be used for micro-assays of various compounds. Some substances which may be determined by a chemiluminescence method are listed below:

CHEMILUMINESCENCE ASSAYS (See Ref. 4)

INORGANIC	ORGANIC
IRON, COPPER, COBALT, CADMIUM, VANADIUM, OZONE, CYANIDES, H_2O_2	α-AMINO ACIDS, ATP, OXIMES, PHENOLES, ORGANOPHOSPHORUS COMPOUNDS (TABUN, SARIN)

Many other compounds such as organic peroxides, glucose, vitamin C, nitroaniline, aniline, resorcinol, pyrogallol, methyl-, ethyl- and propylalcohol may be determined by chemiluminescence using the luminol reaction.[20]

Under optimal conditions, the sensitivity of this new analytical method can exceed even activation analysis, which may suggest the importance of chemiluminescence methods for future research. Although, with their highly sensitive photomultipliers and their excellent counting electronics, liquid scintillation counters are most useful measuring devices for these methods, the applicability of these instruments has so far been quite underestimated and unrecognised.

Bioluminescence assay

Following the suggestion of Tal et al.,[21] in our laboratory we have elaborated a routine method for the quantitative microdetermination of adenosine triphosphate (ATP) using the bioluminescence reaction with firefly enzyme and a liquid scintillation counter as a photodetector.[22] With this instrument, the sensitivity of ATP assay is greatly increased and the costs for enzyme are markedly reduced. Essentially similar procedures for ATP determination were developed and extensively studied by Schram[23] and by Stanley and Williams.[24] In July 1970 at the San Francisco LSC Meeting these authors presented a bioluminescent assay for the specific determination of FMN (flavin mononucleotide) and NADH (reduced nicotinamide adenine dinucleotide).[25,26]

In recent years we have done several thousand ATP determinations with a liquid scintillation counter. I am quite sure that other analytical methods based on chemi- or bioluminescence reactions can easily be performed with this instrument. I hope that these enlarged possibilities of application will be included in the design of future liquid scintillation counters and in the experimental methodology of investigators who use counters.

REFERENCES

1. B. W. Fox, Proceedings of the Beckman Summer School, University College, London, England, 1966, p. 1 d.
2. D. A. Kalbhen, Communication, 11th Symposium on Advances in Tracer Methodology, Boston, Mass., USA, 1966.
3. D. A. Kalbhen, *Intern. J. Appl. Radiation Isotopes* **18**, 655 (1967).
4. D. A. Kalbhen, in E. D. Bransome (Ed), *Liquid Scintillation Counting*, Grune & Stratton Inc, New York, 1970, pp. 127 and 337.
5. D. M. Hercules, in E. D. Bransome (Ed), *Liquid Scintillation Counting*, Grune & Stratton Inc, New York, 1970, p. 315.
6. E. D. Bransome and M. F. Grower, in E. D. Bransome (Ed), *Liquid Scintillation Counting*, Grune & Stratton Inc, New York, 1970, p. 342.
7. R. J. Herberg, *Anal. Chem.* **32**, 42 (1960).
8. R. J. Herberg, *Science* **128**, 199 (1958).
9. J. B. Birks, *The Theory and Practice of Scintillation Counting*, Pergamon Press, London, 1964, p. 279.
10. K. D. Gundermann, *Chemilumineszenz organischer Verbindungen*, Springer Verlag, Berlin, 1968.
11. J. Winkelman and G. Slater, *Anal. Biochem.* **20**, 365 (1967).

12   D. L. Hansen and E. T. Bush, *Anal. Biochem.* **18**, 320 (1967).
13   W. M. Walter and A. E. Purcell, *Anal. Biochem.* **16**, 466 (1966).
14   E. A. Shneour, S. Aronoff and M. R. Kish, *Intern. J. Appl. Radiation Isotopes* **13**, 623 (1962).
15   H. M. Fales, *Atomlight* **25**, 8 (1963).
16   D. A. Kalbhen, and A. Rezvani, International Conference on Organic Scintillators and Liquid Scintillation Counting, San Francisco, California, USA, 1970.
17   M. S. Patterson and R. C. Greene, *Anal. Chem.* **37**, 854 (1965).
18   D. T. Mahin and R. T. Lofberg, *Anal. Biochem.* **16**, 500 (1966).
19   A. C. Houtman, *Intern. J. Appl. Radiation and Isotopes* **16**, 65 (1965).
20   R. F. Wassiljew, *Ideen des exakten Wissens*, Vol. 1, Deutsche Verlags-Anstalt, Stuttgart 1968, p. 49.
21   E. Tal., S. Dikstein and F. G. Sulman, *Experientia* **20**, 652 (1964).
22   D. A. Kalbhen and H. J. Koch, *Z. Klin. Chem. Klin. Biochem.* **5**, 299 (1967).
23   E. Schram, in E. D. Bransome (Ed), *Liquid Scintillation Counting*, Grune & Stratton Inc, New York, 1970, p. 129.
24   P. E. Stanley and S. G. Williams, *Anal. Biochem.* **29**, 381 (1969).
25   E. Schram, International Conference on Organic Scintillators and Liquid Scintillation Counting, San Francisco, California, USA, 1970.
26   P. E. Stanley, International Conference on Organic Scintillators and Liquid Scintillation Counting, San Francisco, California, USA, 1970.

DISCUSSION

M. Krichevsky: What is the effect of changing discriminator settings in rejecting chemiluminescent pulses? Also pyruvic acid or salts thereof might be considered as scavengers for peroxides in chemiluminescent situations as α-ketoacids react quantitatively in an oxidative decarboxylation.

D. A. Kalbhen: In answer to your first point, this is possible only for unquenched carbon-14 samples. As to your second point, we have not tried this but it seems to be a good idea for future investigation.

D. S. Glass: Since chemiluminescence is a low energy phenomenon the problem can be at least brought to light using the double ratio technique (see E. T. Bush, *Int. J. Rad. Isotopes* **19**, 1147 (1968) and also D. S. Glass, *Proceedings of the 2nd International Symposium on organic scintillators and liquid scintillation counting*, San Francisco, July 1970 for the application of this method).

D. A. Kalbhen: The channels-ratio technique is an excellent device to detect chemiluminescence if quenching is not too high and discriminators are set for this purpose.

B. W. Fox: I have three points: *(a)* to what extent do metals influence the chemiluminescence from biological samples from dioxane/alkali mixtures? *(b)* referring to the *p*-aminosalycilate chemiluminescence first falling from a maximum and then rising again to a maximum after about 6–7 hr. Have you an explanation for this? *(c)* may not oxygen *and* a metal be necessary for chemiluminescence?

D. A. Kalbhen: *(a)* Metals do influence chemiluminescence intensity. *(b)* The phenomenon may be due to reaction products which may reduce or increase quenching. *(c)* This is true for ATP-bioluminescence dependent upon Mg^{2+} and oxygen.

D. A. Kalbhen: As to your first point, the intensity of chemiluminescence depends mainly on the peroxide content. In answer to your second: Yes, different scintillators have effects of light output or transmittance (or wave length shifting) so there are different counts for the same chemiluminescence reactions.

B. Scales: Does the chemiluminescence intensity correlate with different scintillators or is it only a function of the solvents and the pH used? Also, have you examined the chemiluminescence intensity of one solvent sample using different scintillators?

C. P. Bond: Have you ever tried using stannous chloride to 'kill' chemiluminescence as advocated by Beckman Instruments? We have found that this virtually eliminates chemiluminescence at least under some circumstances.

D. A. Kalbhen: We have not yet tried this.

J. H. Bates: I presume you carry out mixing of solutions in the absence of light to preclude phosphorescence effects?

D. A. Kalbhen: Yes, but the reduced light of an osmium bulb does not produce significant phosphorescence.

J. A. B. Gibson: How do you allow for the time dependence of the chemiluminescence in ATP estimation?

D. A. Kalbhen: It is advisable (and most users do) to start light measurement exactly 30 s after mixing the reagent and sample, i.e. the enzyme and ATP.

L. Schutte: When measuring chemiluminescence, did you use the coincidence technique? One might expect a higher efficiency when using single photomultipliers.

D. A. Kalbhen: We used *in* and *off* coincidence circuitry. Since chemiluminescence is a *single* photon event, you get much higher pulse rates when coincidence is *off*.

R. Evans: Did you use standardised illumination conditions during the preparation of the various cocktails for chemiluminescence measurement?

D. A. Kalbhen: Yes, it was necessary to avoid lamps which emit u.v. light. We used the reduced light of osmium bulbs.

B. Legg: Is it not possible to overcome the problem of chemiluminescence of peroxides in strong base simply by the addition of a reducing agent or by heating after the decolourising step in the sample preparation?

D. A. Kalbhen: This is possible, but will not completely eliminate chemiluminescence in most cases (see Ref. 11 to this paper).

W. Kolbe: Is there also an influence of temperature on the decay time of chemiluminescence reaction?

D. A. Kalbhen: Yes. The decay is faster at higher temperatures, so it is advisable to store samples obtained at 20–70° for decay.

J. B. Birks: (i) The fluorescence lifetime of liquid toluene of the highest commercial purity is 26 ns but the removal of residual peroxides increases this to 39 ns. (ii) Although spurious counts due to chemiluminescence can be eliminated by allowing the reaction to proceed to completion, the reaction by-products will quench the scintillation efficiency.

D. A. Kalbhen: (i) This may be due to catalytic reactions of contaminants. (ii) Very little is known about reaction intermediates or end-products occurring during chemiluminescence. An increase as well as a decrease in quenching has been observed.

Chapter 2

The Absolute Method of Measurement of Carbon–14 Activity by 4π Liquid Scintillation Counter

T. Radoszewski

The Institute of Nuclear Research, Warsaw, Poland

INTRODUCTION

Standardisation of radionuclides by the use of the 4π liquid scintillation counter is possible due to the full geometry of counting and due to the absence of the main counting corrections such as self-absorption in the source and absorption in the source holder. In the case of carbon-14, as a low energy beta emitter, the so called 'cut off energy' of the counter should be taken into account. It determines the lowest energy of the beta particle which is needed to be counted by the scintillation counter. Here the results of measurements of n-hexadecane labelled with carbon-14 are given, which was counted in single channels, in the coincidence and in the parallel systems.

THE COUNTING EQUIPMENT

All measurements have been made in the home-made scintillation head shown in Fig. 1. It contains the glass vial (12 ml in volume), a light pipe made of perspex and two photomultipliers (EMI 9514S).

The electronic equipment shown in Fig. 2 enabled counts to be made in the single channels, in the coincidence, and in the parallel system (as a sum) with variable levels of discrimination (from 35 to 145 mV, before amplification). It contains two cathode followers, two discriminators, two amplifiers (with a fixed gain of amplification 275 V/V) the coincidence unit (type A 491-A 422 Rochar Electronique) with a resolving time of 1 μs, the sum unit, the quenching unit with a regulated dead time from 5 to 150 μs and the scaler.

PPO in toluene (3 g/l) and PBD in toluene (10 g/l) have been used as scintillators. The normal volume of scintillator used in all counting was 5 ml. All measurements were carried out with the scintillation head placed in a refrigerator at a temperature of -20°C.

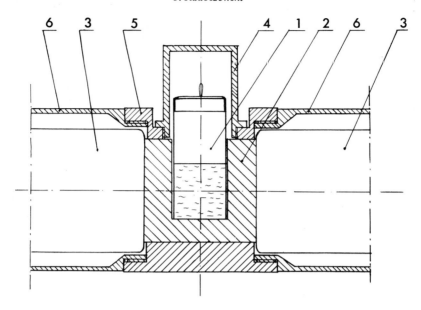

Fig. 1: The double scintillation head. 1. Counting vials with scintillator and a sample. 2. Light pipe made of perspex. 3. Photomultipliers. 4. Sleeve covering the counting vial. 5, 6. Body of the scintillation head, shielding the light pipe and photomultipliers.

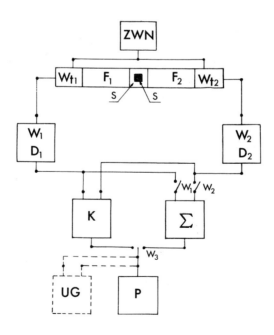

Fig. 2: Block diagram of the counting equipment. ZWN: E.h.t. unit. F: Photomultipliers. W_t: Cathode followers. W, D: Discriminator with amplifier. K: Coincidence unit. Σ: Sum unit. P: Scaler. U.G.: Quenching unit.

THE COUNTING METHOD

Several counting vials with 5 ml of scintillator in each and a small amount of n-hexadecane labelled with carbon-14 (supplied by The Radiochemical Centre, Amersham) were prepared. The quantity of the radioactive material in each sample was determined by weighing. First, the anode characteristics were checked to choose the proper e.h.t. value as the working parameter.[1] The anode characteristics of the separate channel showed some displacement owing to the different sensitivities of the photomultipliers. In spite of this the anode characteristics for the parallel system and for coincidence were taken for equal e.h.t. value in the channels. The characteristics shown in Fig. 3 were taken for a discrimination level of 35 mV and for two different values of the dead time (for the single channel and the parallel system), i.e. 5 and 50 μs.

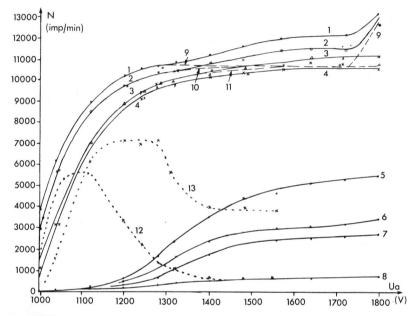

Fig. 3: The anode characteristics of the 4π liquid scintillation counter for 35 mV of discrimination level. 1, 2, 3, 4: Characteristics (after subtracting background) for the parallel system, channel I, channel II and the coincidence system. 5, 6, 7, 8: Background characteristics for the parallel system, channel I, channel II and for the coincidence system. 9, 10, 11: Characteristics for the parallel system, channel I and channel II with the quenching unit of 50 μs dead time. 12, 13: The counting rate of a sample to background ratio $\frac{N^2}{N_C}$ for the parallel and coincidence system.

To show the influence of e.h.t. value on the results of counting, the discrimination characteristics were taken for coincidence and parallel systems for two different values of e.h.t. and are shown in Fig. 4. The value of 1320 V was chosen and fixed for further counting because it ensured the highest possible counting efficiency.

The counting rate of the background from the scintillation was, of course, higher in the parallel system than in the remaining systems and it was kept at a level of 2000 counts/min by the low temperature of the refrigerator ($-20°C$).

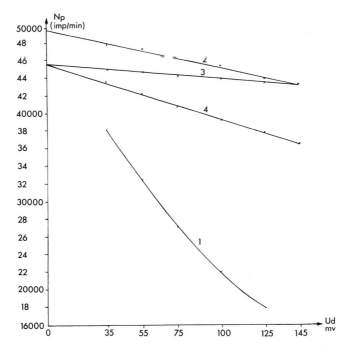

Fig. 4: The discrimination characteristics of the 4π counter. 1, 2: Curves for the parallel system for anode voltage 1120 V and 1320 V correspondingly. 3, 4: Curves for the coincidence system for anode voltage 1320 V and 1560 V correspondingly.

All sources were counted with a variable discrimination level (from 35 to 145 mV) to obtain the discrimination curves and to extrapolate the counting rate to the zero level of discrimination.

No counting corrections were applied for the calculation of the radioactive concentration of the counting n-hexadecane except for the correction due to the background of the counter and for the dead time.

RESULTS OF COUNTING

To measure the radioactive concentration of n-hexadecane, 7 samples were prepared and counted in the manner described previously, using the parallel system. The results are shown in Table 1.

Table 1. The results of radioactive concentration for n-hexadecane labelled with carbon-14, counted in the parallel system

No. of sample	Mass of sample (g)	Scintillator	Radioactive Concentration (μCi/g)	
1	0.02203	PBD	1.075	Mean value
2	0.02155	PBD	1.079	
3	0.02178	PPO	1.062	1.072 μCi/g
4	0.02133	PBD	1.091	standard error
5	0.02106	PPO	1.060	
6	0.02144	PBD	1.083	± 0.5%
7	0.01712	PPO	1.058	

The obtained mean value for the radioactive concentration was

$$1.072 \pm 0.005 \, \mu Ci/g$$

The figure given by the Radiochemical Centre was

$$1.07 \, \mu Ci/g$$

Apart from counting in parallel systems, 4 samples were prepared and counted in all three systems to compare the counting efficiency. The results are given in Table 2.

Table 2. Comparison of the counting efficiency for the parallel, coincidence and the single channel systems for n-hexadecane labelled with carbon-14 (the results in the parallel system are taken as 100%).
Scintillator—PBD in toluene.

No. of sample	Parallel system μCi/g	Coincidence system efficiency %	Channel I efficiency %	Channel II efficiency %
1	1.059 mean value	93.6 mean value	95.6 mean value	96.2 mean value
2	1.087	93.4	97.0	97.9
3	1.084 1.073 μCi/g	92.8 93.4	96.3 96.2	96.9 97.0
4	1.061	93.7	95.9	98.8

Compared to the parallel system, the obtained mean value of the counting efficiency was 93.4% for the coincidence system and 96.2%; 97.0% for the two single channels.

THE SECONDARY PULSES

The anode characteristics from Fig. 3 showed some discrepancy in the case of two different dead time values ($5 \mu s$ and $50 \mu s$). This discrepancy can be explained by secondary pulses arising at the scintillation counter after the main pulses. It should be pointed out that the secondary pulses are observed for the single channel and the parallel system counting but not for a coincidence. The time distribution function of the quantity of these pulses is demonstrated in Fig. 5. This time distribution function has been measured by the electronic equipment (shown in Fig. 2) using a coincidence unit with the addition of a delay unit in one channel. The principle of measurement depended upon coincidence cases between the main (true) pulse in one channel and the secondary pulses, arising after the main pulse, in the second channel, with a variable time of delay. As it is shown in Fig. 5, there is a constant counting rate of the 'true' coincidences as a function of the delay time from 0 to $0.8 \mu s$. This is due to the resolving time of the coincidence unit, which amounts to $1.05 \mu s$ and the minimum delay time of the delay unit which amounts to $0.25 \mu s$. The counting rate rapidly decreased for a further delay and increased again

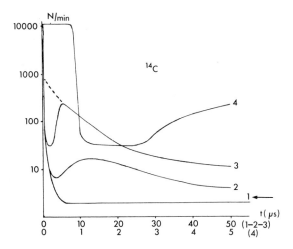

Fig. 5: Time distribution of secondary pulses. 1: Calculated level of the random coincidences between main ('true') pulses. 2: Time distribution for anode voltage 1400 V. 3, 4: Time distribution for anode voltage 2000 V for a delay of 50 μs (3) and 5 μs (4).

after exceeding 3μs of delay time. This is due to the dead time in the channels which amounts to about 3μs. Thus the optimum counting rate has been obtained which gives a rather false picture of the phenomena. One may conclude that the true time distribution should be extrapolated to the zero level of delay and in any case the period from 0 to 5μs cannot be properly investigated by this method. For a period from 5 to 50μs the counting rate is reduced, but even for 50μs it does not obtain the level of a random coincidence rate.

The random coincidence rate has been calculated from the equation:

$$N_R = 2\tau_R N_1 N_2$$

where τ_R is the resolving time, N_1, N_2 are the counting rates in the channel 1 and channel 2.

Taking this into account one can conclude that the curves 2, 3, 4 (Fig. 5) show the time distribution of secondary pulses. It should be emphasised that the amount of secondary pulses does depend upon the e.h.t. value. Further experiments which were carried out, for other radionuclides, lead to the conclusion that the amount of secondary pulses depends also on the energy of β-particles. For example for phosphorus-32 more secondary pulses were observed than for carbon-14.

The reasonable explanation of the origin of the secondary pulses, with time distribution longer than 50μs, is rather difficult. First of all, it should be observed that the secondary pulses arise for a total gain of amplification (photomultiplier + amplifier) high enough to count the single electron from the photocathode.

The second conclusion is that such a long period in time distribution excludes the origin of secondary pulses, as a result of imperfect vacuum in the photomultiplier tube, as the ions from the rest of the gases in the tube should arrive at the electron in a period comparable to 1μs. Providing that the electronic circuit does not cause the secondary pulses (as proved for the circuit in question), the only explanation which the author can suggest is the phosphorescence effect of light from the transparent material such as perspex used as a light pipe, glass of the counting vial and especially the glass envelope of the photomultiplier tube.

CONCLUSIONS

The main conclusion which can be made from the experiments described above is that the 4π liquid scintillation counter can be used as the absolute method for standardisation of carbon-14 compounds. Full efficiency of counting (100%) has been obtained in a limit of the counting errors. According to Horrocks and Studier,[2] the theoretical counting efficiency can be calculated providing that the minimum energy needed for producing one single electron from the photocathode is given. Taking carbon-14, they obtained 99.1% for single channel and 97.1% for coincidence, providing 1 keV as a minimum energy needed and 98.1% for single channel and 93.9% for coincidence providing 2 keV as a 'cut-off energy'. Taking the liquid scintillator efficiency as 70% to anthracene (with 4% of absolute efficiency) and 15% efficiency for photocathode, one may obtain about 620 eV as a minimum energy needed for one single electron, providing that all photons produced by the scintillator strike the photocathode (i.e. full light collection). Taking into account that full light efficiency is never obtained in the scintillation head, but the highest possible efficiency is obviously in the parallel system of counting, it is worthwhile to compare the theoretical data with the obtained results. Concerning the secondary pulses, simply demonstrated in the paper as a possible source of counting errors, one can overcome this problem by applying the quenching unit at high total gain of amplification, or simply by counting in the coincidence system.

REFERENCES

1   T. Radoszewski, *Nukleonika* **5**, 361 (1960).
2   D. H. Horrocks and M. H. Studier, *Anal. Chem.* **33**, 615 (1961).
3   J. B. Birks, *The Theory and Practice of Scintillation Counting,* Pergamon Press, London, 1964.
4   T. Radoszewski, *Nukleonika* **7**, Part 7, to 8 (1962).

DISCUSSION

J. B. Birks: The most probable origin of the secondary pulses is the diffusion-controlled interaction of pairs of triplet-excited solute molecules, leading to the delayed scintillation component. This component is not readily observed in liquid solutions containing oxygen, but the use of low temperature solutions reduces the oxygen solubility and thus reduces oxygen quenching of the triplet states. Your results are similar to those of Voltz and Laustriat at Strasbourg.

T. Radoszewski: I agree with you but we have observed very similar effects when we illuminated just the photomultiplier with light pulses from the light source.

J. A. B. Gibson: The theoretical efficiency at zero bias is always less than 100% and quenching will reduce this efficiency still further. The correction is of the order of 1 or 2% for carbon-14 but for lower energy β-emitters this correction will increase to 30% or more (for tritium, etc.). This correction may be obtained as outlined by J. A. B. Gibson and H. J. Gale, *J. Sci. Instrum. (Physics E)*, Ser. 2, **1**, 99 (1968) (See Fig. 6) and V. Kolarov, Y. Le Gallic and R. Vatin, *Intern. J. Appl. Radiation Isotopes* **21**, 443 (1970).

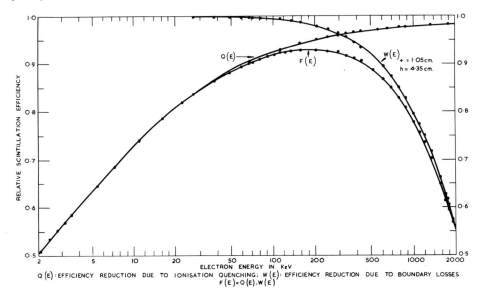

Fig. 6: Variation of scintillation efficiency with electron energy. [From J. A. B. Gibson and H. J. Gale, *J. Sci. Instrumen. (Physics E)* Ser. 2, **1**, 99 (1968)]

T. Radoszewski: As I said, the theoretical efficiency for carbon-14 has been calculated and it amounts to about 99%; therefore the practical full efficiency of counting could be obtained in a limit of error.

D. A. Kalbhen: Are secondary pulses different with other scintillator solutes than PPO and POPOP, or other glass vials or other optical coupling materials?

T. Radoszewski: I did not find any difference between PPO and PBD in toluene.

Chapter 3

Emission Spectra of Liquid Organic Scintillators

E. Langenscheidt

Institut für Physikalische Chemie
Kernforschungsanlage Jülich GmbH, West Germany

INTRODUCTION

Modern scintillation materials are of great importance in applied nuclear physics. For special purposes, the liquid organic scintillant materials exhibit remarkable properties that have also been studied in our laboratory.[1-6] The efficiency of a scintillation system is dependent on all the stages involved in the transfer process of the incident energy (excitation by particles or radiation) in production of light and its conversion into photo-electrons. The light yield of the scintillator is offset by competing non-radiative degradation processes, which are related to the specific energy loss processes in the scintillator. Unfortunately we are lacking a thorough understanding of all phenomena in the scintillant material up till now because the systems are very complex. One can consider three sources of experimental information in the liquid organic counting field: (1) light yield, (2) time dependence of the light pulses, (3) optical spectra. In this work, we consider the optical luminescence spectra of some common scintillation systems, comparing them when using different modes of excitation (α-, β-, γ-rays and ultraviolet light). Some experiments comparing radioluminescence and photoluminescence spectra (mostly excited by β-rays and ultraviolet light) have already been performed, showing the identity of the spectral output in all cases,[7-9] but the sensitivity and resolution of the spectra were insufficient, so that new measurements with more refined experimental conditions were necessary. No detailed discussion was found hitherto concerning the independence of luminescence spectra on the kind of excitation. The study of the luminescence mechanism in detail shows that the behaviour of an excited molecule in the system is affected by local effects of the penetrating radiation (heating, dissociation, electric fields, mutual quenching by neighbouring molecules). The molecular absorption and emission characteristics may be changed therefore when using different radiation. This is plausible because: (1) the scintillation efficiency depends strongly on the type of particle, (2) the shape of a scintillation pulse is affected by the ionisation properties of the incident particle, (3) the luminescence spectra of a given molecule are dependent on the kind of solvent.

THEORY

The solvent of a liquid scintillation system is the constituent absorbing most of the incident energy. For high-energy particles, only the organic aromatic systems are suitable for effectively converting the absorbed energy into light. In the case of scintillator molecules dissolved in a liquid organic medium there is a fundamental influence of the solvent on the spectral characteristics of these molecules. The influence of neighbouring molecules in a liquid medium has not been described in a general form. The results of many investigations are also contradictory. The model commonly utilised to represent the collective actions of the system particle + medium is the Onsager model, in which the molecules are regarded as dipoles in a dielectric medium and exposed to the influence of the internal field of the solvent.

It can be shown theoretically and experimentally that there is a red shift of the absorption and fluorescence spectra of substances in solution as compared with these spectra in the vapour phase.[10-12]

The interaction of the fluorescing molecule with the molecules of the solvent is described by the total energy U arising between two molecules f and s with the dipole moments μ_f and μ_s and the polarisabilities α_f and α_s:

$$U = -\frac{1}{R^6}\left[\frac{2}{3}\frac{\mu_f^2 \mu_s^2}{kT} + \mu_f^2 \alpha_s + \mu_s^2 \alpha_f + \frac{2}{3h}\sum_{\chi,\tau}\frac{a_{\psi\chi}^2 \cdot a_{\sigma\tau}^2}{\nu_{\psi\chi} + \nu_{\sigma\tau}}\right] \quad (1)$$

Σ = sum extended over all transition frequencies $\nu_{\psi\chi}$ and $\nu_{\sigma\tau}$ of both the fluorescing molecule (f) and one neighbouring molecule of the solvent (s) and their corresponding moments of transition $a_{\psi\chi}$ and $a_{\sigma\tau}$.

The ground states of both molecules are denoted by ψ and σ. The four terms of equation (1) mean: 1. the orientation effect, 2, 3. the induction effect, 4. the dispersion effect.

This equation can describe, despite the simplicity of the model used, the shift of the spectra in all cases (dipole or dipoleless molecules in a polar or a nonpolar medium). One can also show that it follows for the total amount of the shift of the spectra of solutions relative to the spectra of vapour:

$$\Delta\nu = \nu_{sol} - \nu_{vap} = \frac{1}{R^3}\left[C_1\frac{2\epsilon - 2}{2\epsilon + 1} + (C_2 + aC_3)\frac{2n^2 - 2}{2n^2 + 1}\right] \quad (2)$$

R = radius of the molecule ($\sim$ effective Onsager radius)
ϵ = dielectric constant ⎫ of the solvent
n = index of refraction ⎭
a = oscillator strength of the electronic transition
C_1, C_2 = functions of the permanent dipole moment of the molecule in the lower and upper electronic states
C_3 = function of the frequency of the purely electronic transition

The magnitude of the red shift is of the order of 4% and can be proved experimentally.[13]

These considerations are valid only for photoluminescence in the first instance. When regarding the interaction with ionising radiation, one must notice the following facts: it is known that the light yield per unit energy absorbed by a liquid scintillator depends on the particle type of the penetrating radiation (for example 250 to 350 eV/photon in the case of α-particles, 25 to 35 eV/photon for β-particles). This fact can be discussed, when regarding the strong electric fields in the neighbourhood of the incident particles, which may produce perturbation of the electronic molecular orbitals. Wright[14] attributes the existence of energy losses by radiationless transitions to these fields.

The incident particle dissipates its energy mainly in producing ionised and excited molecules of the solvent and free electrons along its path. The excited molecules are very rapidly quenched as a result of the strong local electric fields, heating and high density of excitation which exist along the incident particle track immediately after its passage.

There may be created photons in the very first step of interaction whose total number and spectral distribution depend upon the distance from the path of the particle and its velocity. These primary photons would have indeed a modified spectral intensity distribution because the dipole moments of the molecules near the particle track may be altered. Also an effect analogous to the already mentioned dispersion phenomenon is possible (see equation (1)). So one should expect a difference between radioluminescence and photoluminescence spectra if these primary photons contribute to the total luminescence. But if the spectra do not change with the mode of excitation, one has to draw the following conclusions:

(a) Most of the primary photons are reabsorbed within the volume element in question, producing light perhaps in later stages of the process.

(b) The main portion of fluorescent light is observed, when the strong quenching fields due to the incident radiation no longer have a considerable influence, so that no distortion of emission spectra is caused. It follows that detectable luminescence is emitted from excited neutral molecules resulting from recombination phenomena.

It is shown experimentally that the optical spectra of several luminescent organic systems are indeed independent of the mode of excitation within the experimental error.

To account for the observed differences in the light yield of α- or β-excitation, one has to regard further the secondary processes such as intermolecular light transfer and fluorescence with their inevitable degradation and energy loss mechanisms.

The above mentioned most important primary processes are not very well understood. The excitation and ionisation densities are dependent on the velocities of the incident particles and change along their track. The ions and electrons recombine during a short time (10^{-11} to 10^{-13} s).

The decay times τ for light pulses of liquid scintillators are known to be 10^{-8} to 10^{-10} s. τ consists of a fast and a slow component; the corresponding times differ by a factor of about 100. The fast component includes decaying singlet states, but the remainder also a rather large number of excited molecules in triplet states. The excited triplet state is produced mainly by recombination processes, the singlet state by direct excitation. In the case of proton- or α-excitation, the density not only of excited molecules, but also of ions and dissociation products is greater than in the case of β or ultraviolet excitation. It follows that with greater specific ionisation more of the excited triplet state is produced, so that the decay times of the slow pulse components will increase.[15] In addition, heavily ionising particles will cause intense electric quenching fields, especially near the end of the range. During this period, little light may be produced.

The strong quenching effect in the immediate vicinity of the particle track lasts for 10^{-10} s, being terminated as a result of rapid migration of the excitation away from the particle track. So the ions formed along the path of the incident particle are relatively unaffected by the quenching conditions and cause a new crop of excited molecules. It follows that the light yield and emission spectrum of the slow component should be nearly independent of the mode of excitation, while the intensity of the fast component will be less in the case of alpha induced emission than for beta excitation. On the other hand, a modified fluorescence spectrum should be correlated to this fast component.

Further discussion has to regard the relative roles of excited and of ionised molecules in detail (including free radicals) and to study the slowly decaying components of the scintillation when one influences the recombination process by external fields or by electron attaching molecules.

Pulse rise, time measurements[16] as well as other experimental methods give weight to the energy exchange theory (Kallman-Furst) of liquids. Energy transfer in solutions from the lowest excited states (triplets and singlets) of donator molecules to an excited state of the acceptor solute molecules (lowest singlet or triplet) and the consequent fluorescence of these solute species should not be influenced by the strong primary quenching centers for the reason of the short time of the strong quenching effect (10^{-10} s). In references 17 to 22 reasons are discussed for identical mechanisms of energy transfer in all possible modes of excitation.

EXPERIMENTAL

Luminescence spectra were obtained with the aid of a modified Cary 14 recording spectrophotometer (dispersion 40 Å/mm in the visible range). The output of the multiplier phototube was fed to a d.c. amplifier (vibrating reed electrometer) and a pen recorder. It was therefore possible to measure the very small currents due to the faint light intensities (about 2×10^{-4} 1m) of the liquid scintillator excited by α-, β- or γ-rays. The reasons for the small light output of the liquid scintillation detector were: (1) relatively weak activity of the radioactive sources used to avoid effects of radiation damage to the organic system, (2) strong transmission losses due to the optical properties of the spectrograph (poor illumination of the slit, although a condensing lens of 5.0 cm focal length was used to increase the entrance aperture; reflection losses from surfaces and absorption within the optical elements).

The multiplier tube had to be cooled to -25 or -80°C to reduce the dark current (Fig. 1).

Ultraviolet Excitation

To excite the scintillation solution with ultraviolet light, the Cary fluorescence attachment model 1412 was used. The exciting light consisted mainly of lines in the region of 2537 and 3700 Å. Rectangular cells of 0.5, 1.0 and 10 mm optical path lengths were filled with the solutions under investigation.

When the exciting light is strongly absorbed, it makes a great difference whether the

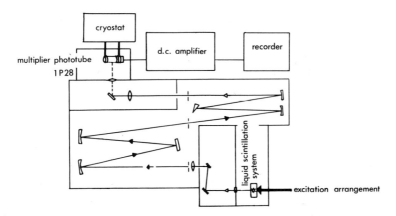

Fig. 1: Experimental arrangement.

fluorescence is observed from the front or the back of a thick liquid layer, because more or less of the shortwave part of the emitted light may be removed by reabsorption. This effect was studied by the use of different path lengths for the u.v. cells and various concentrations of solute molecules in the organic liquid. Further absorption spectra were run.

Excitation by α-particles, where the penetration is small, makes front-surface observation combined with small path lengths desirable, β- and γ-ray excitation with deeper penetration will cause smaller reabsorption effects. Hence it is necessary to plan cautiously experimental conditions for comparing studies of photofluorescence and scintillation spectra by preceding studies of self-absorption effects which depend mainly on exciting radiation, physical dimensions of cells used and on the absorption spectrum of the system.

α-Excitation

The α-source (Fig. 2) consisted of 1 mCi americium-241 mounted on a platinum strip and plunged in the solution opposite to the entrance slit. Because of the low α-range in liquids (5×10^{-3} cm for 5 MeV α-particles), part of the optical luminescence spectrum is modified by the reabsorption phenomena.

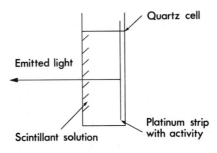

Fig. 2: α-Excitation.

β-Excitation

For the purpose of β-excitation of the scintillator, 0.5 mCi promethium-147 was used in the same manner as in the case of α-excitation. But now, self-absorption effects are smaller, because the β-range is larger (0.1 mm for 0.2 MeV β-particles).

γ-Excitation

A 2 Ci iridium-192 source was applied for excitation of the scintillant material with γ-rays. Self absorption of the obtained spectra is still smaller than in the two cases above, because contrary to α- or β-excitation, each volume element of the scintillator is emitting luminescence radiation when excited in this way.

The following parameters may influence the shape of the technical fluorescence spectrum:
(a) Re-absorption of the emitted photons, secondary fluorescence. *(b)* Concentration of the scintillant molecules. *(c)* Formation of dimers. *(d)* Impurities in the system, quenching agents. *(e)* Temperature. *(f)* Over-all spectral sensitivity of the spectrograph.

(a) This important point has already been discussed. If the optical path lengths of the cells are carefully selected and the concentrations of the solutes are rather low, only the short wavelength-part of the emission spectra obtained may be altered so that corresponding differences in peak-height when changing the type of exciting radiation will not be regarded as fundamental for the scintillation mechanism.

(b), (c) Concentration of the scintillant solutes was changed in all modes of excitation systematically to investigate the amount of re-absorption of a given system. When discussing the structure of the spectra from different modes of excitation, the influence of the slit width setting of the spectrophotometer on the resolution of the spectra was taken into consideration.

The usual emission spectrum, characteristic of the monomeric electronic π-singlet states is absent under intense electron bombardment. Under these conditions the structureless emission observed is due to excimers, resulting from ion recombination processes. Intensities of the exciting radiation used in our work were too small to cause strong monomer quenching, so we could neglect the influence of dimers.

(d), (e) An influence of additional impurities in the scintillant solutions was not found to modify seriously the spectral intensity distribution. The systems were free of oxygen, which causes only diminution of the light yield, not a distortion of the spectral emission characteristics.

(e) It can be shown[23] that temperature effects may be ignored in our case.

(f) The spectral sensitivity of the dispersing system, especially the response curve of the photomultiplier 1 P 28 was determined and compared with data of the manufacturer (RCA). This was realised by application of a fluorescent solution as a quantum counter.[24] In the wavelength region of about 2000 to 3500 Å, a 10^{-2} M solution of 1-dimethyl-aminonaphthalene-7-sodium-sulphonate in water was used. For 3500 to 5900 Å a 1.25×10^{-2} M solution of Rhodamine B in ethylene glycol was used.

With complete absorption of the exciting light at all wavelengths λ, the observed fluorescence intensity is independent of λ (if the fluorescence efficiency is constant), of concentration and of the extinction coefficient.

The spectral response $\eta(\lambda)$ of the multiplier phototube is calculated from

$$\eta(\lambda) = k \times \frac{I(\lambda)}{I_o(\lambda)} \qquad (3)$$

$I(\lambda)$ multiplier phototube output without the quantum counter

$\frac{I_o(\lambda)}{k}$ multiplier phototube current when the fluorescing solution is used

k normalised to $\eta_{max} = 1$

The observed variation of η with λ agrees closely with the RCA data of the S-5 spectral response characteristics (η_{max} at 3400 Å, see Table 1).

Cooling of the tube did not affect its response curve,[25] but when using γ-radiation in our experiments, the tube had to be carefully shielded and only cooled to -30°C, as scintillation of the quartz window of the photomultiplier due to penetrating γ-rays decreased when cooling was reduced.

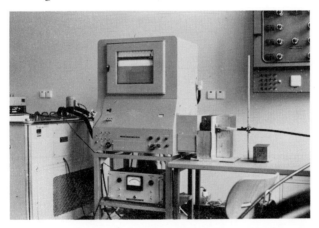

Fig. 3: Experimental arrangement for the γ-excitation experiment.

Table 1. Spectral Sensitivity $\eta(\lambda)$ of the Multiplier Phototube 1P28 (RCA)

$\lambda(\text{Å})$	$\eta(\lambda)$	$\lambda(\text{Å})$	$\eta(\lambda)$
2100	0.377	4050	0.828
2200	0.420	4100	0.820
2250	0.438	4150	0.811
2300	0.447	4200	0.811
2350	0.605	4250	0.786
2400	0.532	4300	0.794
2450	0.589	4350	0.780
2500	0.645	4400	0.777
2550	0.802	4450	0.721
2600	0.831	4500	0.698
2650	0.756	4550	0.698
2700	0.780	4600	0.721
2750	0.744	4650	0.693
2800	0.919	4700	0.681
2850	0.843	4750	0.648
2900	0.905	4800	0.623
2950	0.936	4850	0.597
3000	0.839	4900	0.572
3050	0.834	4950	0.552
3100	0.878	5000	0.516
3150	0.886	5050	0.488
3200	0.894	5100	0.462
3250	0.981	5150	0.425
3300	0.959	5200	0.404
3350	0.971	5250	0.376
3400	1.000	5300	0.356
3450	0.977	5350	0.328
3500	0.991	5400	0.296
3550	0.967	5450	0.266
3600	0.946	5500	0.238
3650	0.955	5550	0.212
3700	0.972	5600	0.191
3750	0.910	5650	0.181
3800	0.884	5700	0.178
3850	0.878	5750	0.168
3900	0.845	5800	0.157
3950	0.839	5850	0.141
4000	0.839	5900	0.117
		5950	0.099

RESULTS

The emission spectrum of a specimen is a plot of luminescence intensity (quanta per unit wavelength interval) against wavelength. This apparent spectrum may be corrected with three wavelength dependent factors: quantum efficiency of the photomultiplier, band width and transmission factor of the monochromator.

The spectra shown in the Figs. 4 to 12 are not corrected, as variation of the spectral sensitivity in the wavelength region of interest is only $\leqslant 15\%$. Further, the obtained spectra are indeed directly comparable because the same experimental conditions were chosen for a set of spectral intensity curves of a system. The obtained emission spectra of various liquid solutions show a spectral distribution of the luminescence produced independent of any particular method of excitation. However, under radioactive excitation of a liquid system a complete shift of its original spectrum on addition of a second phosphor with an emission spectrum of longer wavelengths is observed; in ultraviolet irradiation all compounds will fluoresce. This may be explained because in the latter case the solute molecules are directly excited, as the ultraviolet excitation energy is not very different from optical molecular levels. When using high energy particles direct excitation is not possible, but there is successive energy transfer from the solvent to the solute molecules fluorescing in the longest wavelength bands.

On the other hand the ratio of fluorescence intensity of the solvent and solute molecules depends on the respective ratio of their concentrations; this ratio is not affected by the mode of excitation.

When comparing the spectra, we ignore their spectral region subject to self-absorption.

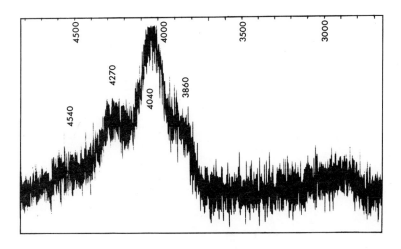

Fig. 4: Anthracene in toluene, excited by γ-radiation.

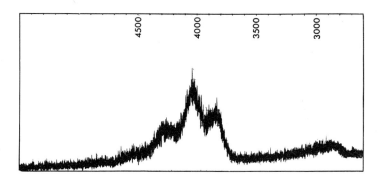

Fig. 5: Anthracene in toluene, excited by β-radiation.

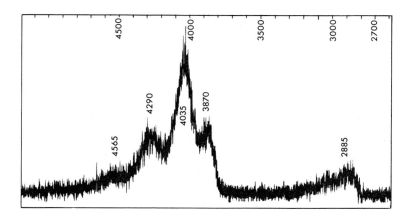

Fig. 6: Anthracene in toluene, excited by α-radiation, effect of self-absorption.

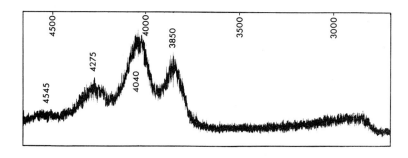

Fig. 7: Anthracene in toluene, excited by α-radiation.

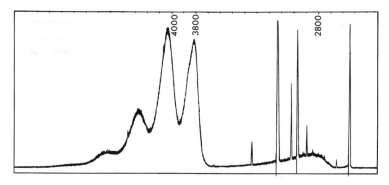

Fig. 8: Anthracene in toluene, excited by ultraviolet light.

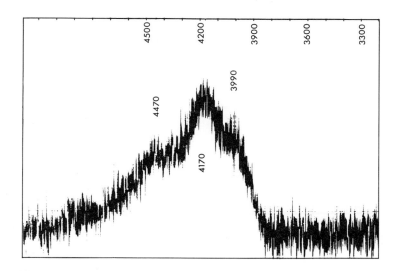

Fig. 9: POPOP in toluene, excited by γ-radiation.

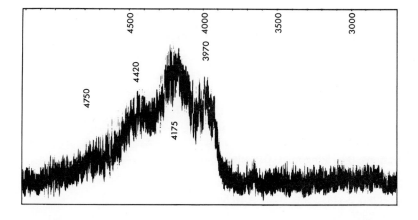

Fig. 10: POPOP in toluene, excited by β-radiation.

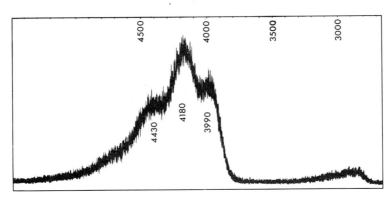

Fig. 11: POPOP in toluene, excited by α-radiation.

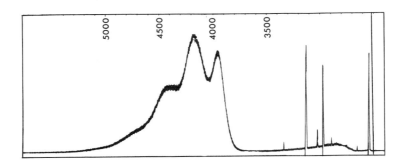

Fig. 12: POPOP in toluene, excited by ultraviolet light.

Emission spectra of the following systems were obtained:
(a) Anthracene in toluene (Figs. 4 to 8) rel. intensities:
emission bands: 3815, 4035, 4270, 4520, 4860 Å (0.91:1.0:0.41:0.11:0.04).
absorption bands: 3135, 3265, 3418, 3595, 3790 Å.
concentrations: 0.334 mg/ml, 0.073 mg/ml.

(b) 1.4-di-[2-(5-phenyloxazolyl] benzene (POPOP) (Figs. 9 to 12) in toluene
emission: 3960, 4180, 4415, 4750 Å (0.875:1.0:0.58:0.2).
absorption: 3500, 3628, 3810 Å.
concentrations: 0.207 mg/ml, 12 mg/ml.

(c) 2-phenyl-5(4-biphenyl)-1, 3-oxazole (PBO) in toluene
emission: 3440, 3610, 3780 Å (0.71:1.0:0.8).
absorption: 3060 Å.
concentrations: 0.075 mg/ml, 0.019 mg/ml, 0.1 mg/ml.

(d) *p*-terphenyl in toluene
emission: 3280, 3425, 3560, 3765 Å (0.84:1.0:0.65:0.27).
absorption: 2826 Å.
concentrations: 0.065 mg/ml, 0.255 mg/ml, 0.02 mg/ml.

(e) *p*-quaterphenyl in toluene
emission: 3540, 3710, 3885 Å (0.97:1.0:0.63).
absorption: 3000 Å.
concentrations: 0.077 mg/ml.

(f) 0.7% PPO, 10% naphthalene in dioxane
PPO = 2,5-diphenyl-1,3 oxazole
emission: 3460, 3630, 3780 Å.
absorption: complex.

Experimental comparison of radioluminescence and photoluminescence indicates the identity of the corresponding emission spectra relative to the spectral intensity of the bands as well as to their spectral position. The estimated tolerances are 7% for intensity and 1% for the wavelength position.

When comparing the emission spectra of the pure solvents, a red shift in the intensity maximum of about 50 Å was observed when changing from ultraviolet-excitation to excitation with γ-rays. This effect of the order of 1% may be explained with self-absorption phenomena.

The independence of the luminescence spectra from the mode of excitation confirms the importance of recombination phenomena in liquids interacting with high energy particles. Detailed information about the phenomena involved is difficult to derive, because complete analysis of all processes cannot be obtained by mere investigations of the optical properties of liquids.

ACKNOWLEDGEMENT

The author wishes to thank Dr. H. Ihle for his devoted help throughout the whole work.

REFERENCES

1. H. Ihle, A. P. Murrenhoff and H. Aulich, Jül–127–PC, 1963.
2. H. Ihle, M. Karayannis and A. P. Murrenhoff, Jül–347–PC, 1966.
3. H. Ihle, M. Karayannis and A. P. Murrenhoff, Jül–298–PC, 1965.
4. H. Ihle, E. Langenscheidt and A. P. Murrenhoff, Jül–491–PC, 1967.
5. H. Ihle, M. Karayannis and A. P. Murrenhoff, International Atomic Energy Agency, Vienna, 1965.
6. H. Ihle, M. Karayannis and A. P. Murrenhoff, *Atompraxis* **14**, 397, (1968).
7. J. B. Birks, C. L. Braga and M. D. Lumb, *Brit. J. Appl. Phys.* **15**, 399, (1964).
8. L. Roth, *Phys. Rev.* **75**, 983, (1949).
9. H. Kallmann and M. Furst, *Phys. Rev.* **81**, 853, (1951).
10. A. S. Cherkasov, *Opt. i Spektroskopiya* **2**, 597, (1957).
11. S. Sambursky and G. Wolfsohn, *Phys. Rev.* **62**, 357, (1942).
12. F. London, *Trans. Faraday Soc.* **33**, 8, (1937).
13. N. G. Bakhshiev, *Bull. Acad. Sci. USSR, Phys. Ser. (English Transl.)* **24**, 593, (1960).
14. G. T. Wright, *Proc. Phys. Soc. (London) B* **68**, 929, (1957).
15. D. L. Horrocks, *Rev. Sci. Instr.* **34**, 1035, (1963).
16. H. Kallmann and G. J. Brucker, *Phys. Rev.* **108**, 1122, (1957).

17   I. B. Berlman, *J. Chem. Phys.* **33**, 1124, (1960).
18   S. Lipsky and M. Burton, *J. Chem. Phys.* **31**, 1221, (1959).
19   F. H. Brown, M. Furst and H. Kallmann, *Discussions Faraday Soc.* **27**, 43, (1959).
20   J. B. Birks and K. N. Kuchela, *Discussions Faraday Soc.* **27**, 57, (1959).
21   M. Furst and H. Kallmann, *Phys. Rev.* **94**, 503, (1954).
22   S. G. Cohen and A. Weinreb, *Proc. Phys. Soc. (London) B* **69**, 593, (1956).
23   L. G. Pikulik, *Bull. Acad. Sci. USSR, Phys. Ser. (English Transl.)* **24**, 578, (1960).
24   J. B. Birks and I. H. Munro, *Brit. J. Appl. Phys.* **12**, 519 (1961).
25   G. G. Kelley and M. Goodrich, *Phys. Rev.* **77**, 138, (1950).

DISCUSSION

J. B. Birks: Christophorou (Oak Ridge) and Joneleit (Giessen) have observed that liquid benzene and toluene excited by intense electron beams have luminescence spectra differing from the photo-excited spectra. It would be of interest to study whether similar differences occur with α-particle excitation. The reason for the difference has been discussed elsewhere. (J. B. Birks, *Chem. Phys. Lett.* 1970).

E. Langenscheidt: In our case of the application of radio nuclides the spectra of the pure solvents seem to be independent of the mode of excitation; but the pure liquids were not studied when excited by very intense α or β radiation.

D. A. Kalbhen: What is the spectrum produced by the Cerenkov effect of higher energy β-particles in water like? Is it well-defined or diffuse?

E. Langenscheidt: The Cerenkov effect spectrum is diffuse, but this problem has not been studied here.

Chapter 4

Scintillations in Liquid Helium

P. B. Dunscombe

Department of Physics, University of Birmingham, England

Scintillations in liquid helium due to the passage of a charged particle were first observed by Simmons and Perkins in 1961.[1] Subsequent investigations have revealed that the intensity of the scintillations as a function of temperature is different from that for scintillators near room temperature and is almost certainly connected with the superfluid nature of liquid helium below the λ point. This temperature effect is discussed in terms of current theories of liquid helium and in the light of the present state of knowledge relating to the scintillation phenomenon.

Both α and β particles have been used to excite the liquid helium and it is significant that the effect of temperature on the scintillations depends on the particle producing the scintillation. The most thorough investigation of α-induced scintillations has been carried out by the University of Virginia group[2] who have observed scintillations down to 0.3°K. Their results are shown in Fig. 1. The scintillation intensity drops steadily from 4.2°K to the λ point (2.18°K) where it drops sharply. Below the λ point the intensity decreases by about 15% to what appears to be a constant value. The decrease from 4.2°K to 2.18°K can be explained by a change in the density of the liquid which causes a decrease in the α-path length and hence an increase in the fraction of scintillation light absorbed on the source holder. The sharp drop at the λ point is probably due to the cessation of internal boiling in the liquid. The superfluid nature of helium below the λ point gives rise to a very large thermal conductivity which prevents enough heat being localised to form a bubble. The fit to the data below the λ point produced by the authors of reference 2 is discussed later. They noted in this experiment that if all the emitted photons were collected then no decrease was observed below the λ point. These observations clearly require an explanation in terms of a mechanism which delays photon emission to a degree that depends on the temperature of the liquid.

A similar investigation of the effect of temperature on the scintillation intensity induced by β-particles gave the results shown[3] in Fig. 2. It can be seen here that the intensity remains constant with temperature apart from a discontinuity at the λ point. As the path length of electrons is about two orders of magnitude greater than that of

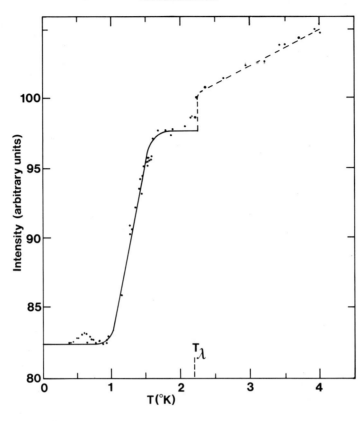

Fig. 1

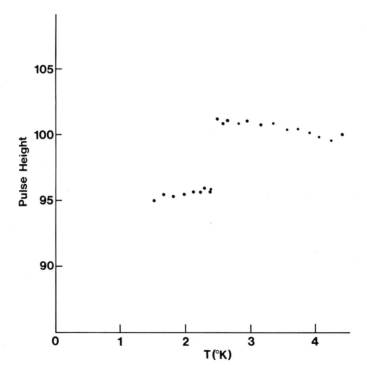

Fig. 2

α-particles no detectable light absorption effects are expected. The discontinuity at the λ point is again probably due to the cessation of internal boiling.

Before attempting to discover the cause of these effects the results of two more experiments which may be relevant should be discussed.

The first experiment[4] was designed to determine the proportion of scintillation light that is due to ion recombination processes. This was carried out by applying an electric field to the region around an α source to draw off the ions and correlate the current with the decrease in scintillation intensity. The conclusion reached here is that about 60% of the light is due to recombination processes. These authors also concluded that the recombination luminescence is not temperature dependent. In the light of reference 2 this conclusion is not justified from their experiment.

The second experiment[5] was designed to determine the spectrum of light emitted from liquid helium following bombardment with electrons. A broad continuum was observed with its maximum centred on 800 Å. The emission is due to the radiative dissociation of the ground vibrational level, $A'\Sigma_u^+$, of He_2. The afterglow period was examined by using a pulsed electron beam but the rather qualitative results make an interpretation difficult. It is possible however, that an effect due to the increased mobility of ions as the temperature is reduced was seen in this experiment. A similar investigation[6] has indicated that the broad peak may, in fact be made up from two peaks at 825 Å and 755 Å.

A spectroscopic study in the longer wavelength region[7] has shown that the metastable state $a^3\Sigma u^+$ is populated at a rapid rate when liquid helium is excited by electrons. The significance of this will become clear.

No studies of the spectrum produced by α-induced excitation have been carried out but there is no reason to expect that the molecular processes will be different from those induced by electrons. In fact the metastable state mentioned above, the $a^3\Sigma u^+$ state, may have been observed in quite a different way by Surko and Reif.[8] Measurements on this neutral excitation indicate that below 0.5°K its lifetime could be quite long (10^{-4} s) but above this temperature the increasing number of thermal excitations in the liquid would cause its rapid decay.

Fischbach et al.[2] have proposed that the effect of temperature on α-induced scintillations is due to the formation of metastable states whose decay rate will depend on the density of excitations (phonons and rotons) in the helium. With this theory they have obtained a good fit to the α-data. The question still remains however, as to why a similar effect is not observed for β-induced scintillations. There is no evidence that the molecular states involved are different for the two methods of excitation. In fact the metastable state most likely to be responsible for an effect of this kind, the $a^3\Sigma u^+$ state, is known to be present in the excitation produced by electrons.

In order to investigate the effect further it will be necessary to know something about the physics of liquid helium.

If the pressure above liquid helium is reduced and the liquid is thus cooled it is found that at 2.18°K (the λ point) a phase transition takes place. This is similar to a second order phase transition but may not rigorously satisfy the requirements for one.

Below the λ point some strange effects occur which give rise to the two fluid model of liquid helium. First, if helium is passed through a narrow channel (10^{-4} cm width) it is found that up to a certain critical velocity the flow-rate is independent of the pressure-head and depends in a complicated way on the width of the channel such that the phenomenon bears no resemblance to the usual concept of viscosity. A disc rotating in liquid helium will however be subject to a frictional drag.

The explanation of these two apparently contradictory observations is that there are two components in helium below the λ point. The super-fluid component is capable of flowing through very narrow channels with little or no drag; and the normal component which is responsible for the viscous effect observed in the rotating disc experiment. Thus the total density is made up from these two components.

$$\rho = \rho_s + \rho_n$$

A theory to explain this effect is based on the assumption that the weakness of the interatomic forces in helium allow it to be treated in a similar way to a gas of bosons. While this theory must have its limitations it does show that below a certain temperature a finite number of atoms will be in a zero energy state.

Another approach to the liquid state is to regard it as a type of solid in which the atoms are free to move from their lattice sites. Landau, using this approach, showed that two types of excitation could exist in the liquid. The phonon corresponds to longitudinal motion while the roton is associated with rotational motion. These phonons and rotons which comprise the normal fluid are embedded in an unexcited background, the superfluid. Thus a calculation of the density of phonons and rotons should yield the relative proportions of the normal and superfluid components.

It has become apparent recently that another type of excitation can exist in helium below the λ point. Excitations of this kind are called vortices and are associated with the superfluid component. Vorticity is a classical idea but in the case of liquid helium it can be shown that the angular momentum of the helium atoms about the core is quantised giving the following relation between the velocity of the atom, v, and its distance from the vortex core, r.

$$v = n \frac{\hbar}{m} \frac{l}{r}$$

where n is an integer and m is the mass of the helium atom. This quantisation arises from the orderliness of the atoms in the superfluid component. The atomic wavefunctions are not perturbed by interactions with phonons and rotons and hence will be single valued for a closed trajectory. These vortices are similar to whirlpools in a stream (for example).

The significance of vorticity in the present problem is that it can trap ions. The mass and size of an ion in liquid helium is however far greater than that of the free ion. The positive ion, which is likely to be He_2^+, gives rise to a polarisation of the atoms in its vicinity. This effectively increases its mass to about 100 helium mass and its radius to about 6.4 Å. The negative ion, which is probably the electron, exists in a 'bubble' in the liquid. Its polarising effect gives it an effective mass of 100 helium masses and a radius of 14.5 Å.

The trapping of ions can be explained using hydrodynamical arguments[9] that involve the reduction in energy when an ion is substituted for a portion of the vortex. In this way lifetimes of the trapped ions have been calculated and it is now known that positive ions escape very easily from vortices at temperatures above about 0.7°K. Negative ions however can remain trapped for times of the order of minutes below 1.6°K and so escape for these can be neglected.

Once an ion is trapped in a vortex line there is less chance of it being scattered by phonons and rotons and hence it can move quickly through the liquid.

The possible connection between vorticity and the scintillation effect was first realised by Agee et al.[10] who carried out experiments to attempt to prove the connection. In their first experiment[10] they produced vorticity by rotating the liquid helium. No effect was observed but as the authors point out the amount of vorticity produced was so small that no effect could reasonably be expected. Their second experiment[11] was designed to produce vorticity by passing a large heat current through the liquid. The effect they observed was explained in a later publication[12] in terms of the effect on the path length of bubbles formed in the liquid by the large heat flow. Thus the connection between vorticity and the scintillation process has neither been established nor refuted.

If any connection between the two does exist it may be possible to explain why the scintillation effects produced by α-particles and electrons are different. The α-particles in the experiments mentioned above originated from a source immersed in the helium and as their path length is only 0.2 mm it is reasonable to expect the liquid to be perturbed in the region of the source. This perturbation would take the form of the production of rotons, phonons and vortices near the source. As the decay time of the vorticity would be of the order of seconds[13] some equilibrium density of vortices would be set up and this would depend on the fraction of superfluid in the liquid, ρ_s/ρ_n, and hence on the temperature.

The situation for an electron however would be quite different as its path length is much longer than that of an alpha particle of similar energy. This would result in a much lower vortex density and any interaction between the vortex and the ions would be small compared to that for ions produced in the α-particle track.

With this high density of vorticity in the α-particle track it is possible that below about 1.7°K the negative ion complex can be trapped in a vortex line and move rapidly out of the region of high ion concentration. This process will inhibit recombination and could give rise to the reduced intensity observed for α induced scintillations.

Another possibility is that the neutral metastable excitation mentioned above is trapped in a vortex and hence protected from collisions with phonons and rotons which would cause its decay. This would give rise to an excitation that would be longer lived in the region of the α source than that in an electron track. There is no direct evidence however for the trapping of these neutral excitations available at the present time.

In conclusion it would seem that there is strong evidence for a connection between the temperature dependence of the scintillation intensity and the production of vortices, although the precise mechanism is still open to speculation. Until more is known about the production of vorticity by an energetic charged particle, it is unlikely that a meaningful fit to the data will be obtained or the mechanism involved exactly understood.

It is proposed to establish if a connection between vorticity and the decrease in scintillation intensity exists by using a low energy β source in the region of the α particle tracks such that ions produced by either particle will be exposed to the same density of vorticity and hence any effect of vorticity on the scintillation intensity should be similar for excitations produced by the two particles.

REFERENCES

1. J. E. Simmons and R. B. Perkins, *Rev. Sci. Instr.* **32**, 1173 (1961).
2. M. R. Fischbach, H. A. Roberts and F. L. Hereford, *Phys. Rev. Letters* **23**, 462 (1969).
3. J. R. Kane, R. T. Siegel and A. Suzuki, *Phys. Letters* **6**, 256 (1963).

4 F. L. Hereford and F. E. Moss, *Phys. Rev.* **141**, 204 (1966).
5 M. Stockton, J. W. Keto and W. A. Fitzsimmons, *Phys. Rev. Letters* **24**, 654 (1970).
6 C. M. Surko, R. E. Packard, G. J. Dick and F. Reif, *Phys. Rev. Letters* **24**, 657 (1970).
7 W. S. Dennis, E. Durbin, Jr., W. A. Fitzsimmons, O. Heybrey and G. K. Walters, *Phys. Rev. Letters* **23**, 1083 (1969).
8 C. M. Surko and F. Reif, *Phys. Rev.* **175**, 229 (1968).
9 P. E. Parks and R. J. Donnelly, *Phys. Rev. Letters* **16**, 45 (1966).
10 Forrest J. Agee, Jr., Robert J. Manning, James S. Vinson and Frank L. Hereford, *Phys. Rev.* **153**, 255 (1967).
11 Frank E. Moss, Frank L. Hereford, Forrest J. Agee and James S. Vinson, *Phys. Rev. Letters* **14**, 813 (1965).
12 James S. Vinson, Forrest J. Agee, Robert J. Manning and Frank L. Hereford, *Phys. Rev.* **168**, 180 (1967).
13 W. F. Vinen, *Proc. Roy. Soc. (London), Ser. A* **242**, 493 (1957).

Chapter 5

A Liquid Helium Polarimeter of Unique Design

J. Birchall,* C. O. Blyth, P. B. Dunscombe,† M. J. Kenny,**
J. S. C. McKee and B. L. Reece

Department of Physics, University of Birmingham, England

INTRODUCTION

Much of the present effort in low energy nuclear physics is directed towards an investigation of spin dependent effects in nuclear interactions and for such an investigation either the incident beam or the target polarisation must be accurately known. The instrument described in this chapter was designed for the precise measurement of neutron beam polarisation in the energy range 2 to 20 MeV. The essential requirements of such a device and the way in which these requirements have influenced the final design are described.

When a beam of neutrons of polarisation P_1 is scattered from a target, more neutrons may be scattered at an angle θ to one side of the original beam direction than at a similar angle to the other. If $L(\theta)$ and $R(\theta)$ are the numbers scattered to the left and right of the beam respectively, then the asymmetry of the scattering is defined as:

$$\epsilon(\theta) = \frac{L(\theta) - R(\theta)}{L(\theta) + R(\theta)}$$

The factor relating this asymmetry to the initial beam polarisation is known as the analysing power of the scatterer, $P_A(\theta)$, and is given by:

$$\epsilon(\theta) = P_A(\theta) P_1$$

$\epsilon(\theta)$ is the experimentally measured quantity, so if it is required to determine P_1 accurately

---

*Now at Universite Laval, Quebec, Canada.
†Paper presented by P. B. Dunscombe.
**Now at AAECRE, Private Mail Bag, Sutherland, N.S.W. 2232, Australia.

$P_A(\theta)$ must satisfy certain conditions. The main conditions are that it should be as large as possible over a reasonable angular range and that it should not change too rapidly with either the energy of the incident neutron or the scattering angle, θ. A further condition is that the scatterer chosen should have a reasonable cross-section for scattering at the angles of highest polarisation so that a statistically accurate measurement of the polarisation may readily be obtained. The nucleus which satisfies these requirements best is the helium-4 nucleus which has no nuclear excited states below 20 MeV.[1]

DESIGN CONSIDERATIONS

Light is emitted from helium when a charged particle is slowed down or stopped in it. Such a charged particle could be a helium nucleus recoiling from collision with an incident neutron. Thus, if the scattered neutron and the recoiling helium nucleus can both be detected one has an efficient method of reducing the background present in neutron experiments. Also if the energy of the recoil can be measured accurately, the background can be further reduced by setting a pulse height window on the recoil spectrum.

The difficulties encountered in obtaining good resolution are due to two factors. The first is that helium scintillates in the far ultraviolet region of the electromagnetic spectrum[2] and these wavelengths are outside the range of conventional photomultipliers. The second problem is that the scintillation light must be transmitted from the liquid helium to the cathode of a photomultiplier which is operating around room temperature. The way in which these problems have been overcome will be described.

The liquid phase of helium was chosen in preference to the gaseous phase for two reasons. The first is the higher density of helium nuclei in the liquid leads to higher counting rates and hence better statistics on the final result. The second reason is that for neutron experiments at energies around 10 MeV in which time-of-flight techniques are employed to measure energies, the rise times of the pulses used in the timing must be rapid, i.e. about 10^{-9} s. Scintillation pulses from liquid helium are known to meet these requirements, whereas those from gaseous helium, even under high pressure, are inferior, particularly as regards decay time. The disadvantage of a liquid helium target is that it needs to be filled regularly,* unlike a gas target which may be run indefinitely without attention. Several high pressure gas targets have already been constructed.[3]

FINAL DESIGN

The liquid helium is contained in a 3" diameter copper bowl, 1/16" thick, at one end of a quartz light pipe. This light pipe transmits the scintillation light to a 56 A.V.P. photomultiplier whose cathode is at room temperature. To convert the scintillation light to a wavelength suitable for detection by the photomultiplier a layer of wavelength shifter is deposited on top of a diffuse reflecting paint which covers the inside of the bowl. It has been found that DPS† is superior to POPOP** as a wavelength shifter in this wavelength region giving an increase in pulse height of about 1.3. The optimum thickness of the layer is 100 $\mu g/cm^2$.

The diffuse reflecting paint is, of course, essential if good resolution is required, otherwise the size of the signal from the photomultiplier for equal scintillations in the

---

*With liquid helium self-filling systems now available commercially, this disadvantage is effectively removed.
†p-p' diphenylstilbene.
**p-bis[2-(5-phenyloxazolyl)] p-benzene.

bowl will depend on position. Another consideration affecting this coat of reflecting paint is the rise time of the light pulse which stimulates the cathode of the photomultiplier. As has been pointed out, this should be as short as possible to reduce the background in the neutron scattering experiment. If the reflectivity of the bowl was 100%, light could be reflected many times before entering the light pipe and then being transmitted to the photomultiplier and this would adversely affect the rise-time of the pulse. Thus a compromise has to be reached between high reflectivity with good resolution and low reflectivity with good timing characteristics. The reflector paint, N.E.560, has been found to be suitable for this purpose.

Because of the large temperature difference between the scintillating medium and the photocathode it is necessary to transmit the light with high efficiency through a medium which will stand the high temperature gradient. Previous designs have used a quartz[4] or sapphire window in the scintillating volume and placed the photomultiplier a short distance away in the insulating vacuum. It was felt however, that improved transmission would be obtained if the photomultiplier was directly coupled through a light pipe to the scintillating volume, thus reducing light losses at additional optical interfaces. The light guide chosen for this purpose is quartz. Not only has quartz a very high transmission, but its cut-off wavelength is well down in the ultraviolet region. In order that the photocathode should remain near room temperature, the light guide is 15" long with its end well above the top plate of the cryostat which acts as a heat sink. The other end of the light pipe is coated with 50 $\mu g/cm^2$ of the wavelength shifter DPS and immersed in the helium as shown in Fig. 1. Its diameter is 2".

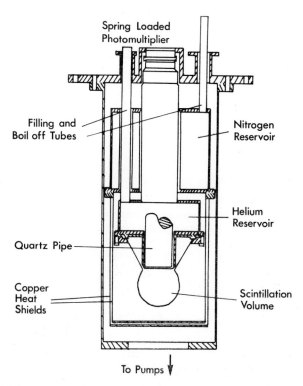

Fig. 1: The polarimeter design.

As the refractive index of helium is 1.026, any light entering the pipe from the helium end is totally internally reflected in the pipe and must leave by the other end if the walls of the light pipe are clean and not in contact with anything. A seal however, has to be made to prevent the escape of helium gas and this takes the form of a thin 'O' ring near the top of the pipe. This total internal reflection condition is also lost when the light pipe is coated at one end with a wavelength shifter, as in this case. Silvering the walls could reduce the light losses but would also increase the variation in transit times up the pipe, hence spoiling both rise and fall times of the detected pulses. It has not been found necessary in the present case to use this technique although it would probably increase the total charge collected at the anode of the photomultiplier.

The light guide is optically coupled to the spring loaded photomultiplier with Dow-Corning C-2-0057 optical coupling grease. A photomultiplier with a quartz window is soon to be used in place of the 56 A.V.P. and it is expected that this will further improve the resolution of the instrument.

The bowl containing the helium is sealed to the helium reservoir with an indium ring. This allows the reflector paint and wavelength shifter to be renewed as necessary. The helium system is supported in the insulating vacuum by three filling and boil-off tubes and is totally surrounded by a copper radiation shield cooled with liquid nitrogen.

The resolution of the instrument has been investigated using α-particles from a polonium-210 source electrodeposited on a silver wire, and helium nuclei recoiling from collision with a neutron (Fig. 2). The results of these tests are consistent with a resolution of 12% for E_α = 9 MeV inclusive of position effects in the scintillating volume and statistical effects at the cathode of the photomultiplier.

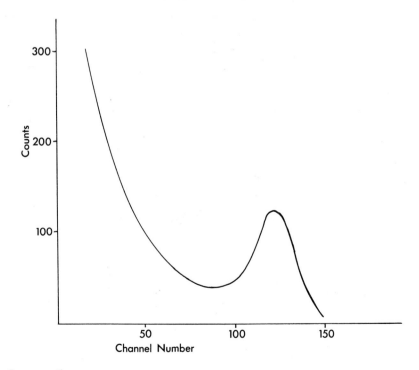

Fig. 2: α-recoil spectrum.

REFERENCES
1. B. Hoop and H. H. Barschall, *Nucl. Phys.* **83**, 65 (1966).
2. J. E. Simmons and R. B. Perkins, *Rev. Sci. Instr.* **32**, 1173 (1961).
3. P. S. Dubbeldam and R. L. Walter, *Nucl. Phys.* **28**, 414 (1961).
4. T. G. Miller, *Nucl. Instr. and Methods* **40**, 93 (1966).

DISCUSSION

J. B. Birks: Does the emission spectrum of liquid helium correspond to that of the excimer (excited dimer) or of the dimer cation (He_2^+)? What is the scintillation decay time?

P. B. Dunscombe: The emission spectrum in the u.v. is largely due to the radiative dissociation of the excited dimer. The He_2^+ ion is formed in the liquid (binding energy about 2 eV). The decay time of the scintillation from liquid helium is of the order of 10 ns.

Chapter 6

The 4π Liquid Scintillation Method for Activity Measurement of Electron Capture Nuclides

Andrzej Tada and Tomasz Radoszewski*

The Institute of Nuclear Research, Warsaw, Poland

INTRODUCTION

The activity of electron-capture nuclides have been previously measured by 4π proportional counter worked under elevated pressure.[1] Hereafter the 4π scintillation counter with plastic scintillators has been used for the same purpose.[2] In both cases the only radiation counted were K-X photons, or partly L-X, as Auger electrons were considerably absorbed at the source holder and at the source itself. Further progress in standardisation of E.C. nuclides relied upon the 4π(x, e)-γ coincidence method. Scintillation counters with a NaI/Tl crystal as well as ionisation chambers were also used with rather poor counting efficiency as Auger electrons were not counted, in these cases, at all.

Here the 4π liquid scintillation method has been used for counting the following E.C. nuclides: chromium-51, manganese-54, iron-55, zinc-65, yttrium-88 and iodine-125. Auger electrons of low energy can be counted in these cases together with K-X photons, and therefore all the advantages of liquid scintillation method for low energy β-emitters counting are valid.

THE COUNTING EQUIPMENT

The counting equipment used for E.C. nuclides counting was the same as for carbon-14 standardisation.[3] Counting could be carried out in parallel and in the coincidence system. The scintillator used was PBD in toluene (8 g PBD/l 1 toluene) and 4 g PPO + 0.5 g POPOP/l 1 toluene with Triton X-100, for some experiments.

THE COUNTING METHOD

Sources to be counted were prepared in the counting vials, by dissolving a small amount of a radioactive solution in the scintillator, together with 1 ml of pure alcohol

---

*Paper presented by T. Radoszewski

(except when Triton X-100 was added). As in the case of β-emitter counting, the anode characteristics for the parallel system and for the coincidence system were checked for each nuclide to choose the proper e.h.t. value as the working parameter.[4] The anode characteristics for iron-55 are shown in Fig. 1. After choosing the proper e.h.t. value, all sources were counted with a variable discrimination level (from 48 to 175 mV) to obtain the discrimination curves and to extrapolate the counting rate to the zero level of

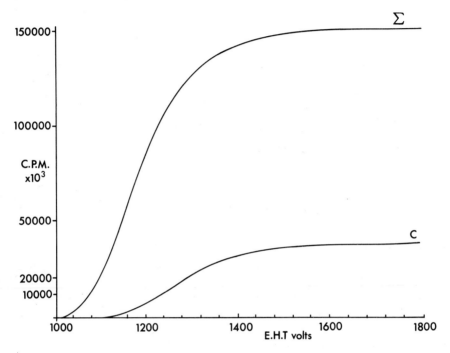

Fig. 1: The anode characteristics for iron-55. Σ: The parallel system. C: The coincidence system.

discrimination. The discrimination curve for iron-55 is shown in Fig. 2. All measurements were made with the additional quenching unit with 48 μs of dead time. The counting temperature was -20°C.

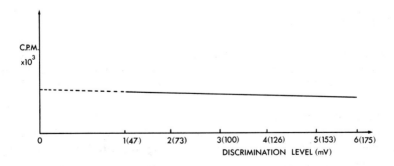

Fig. 2: The discrimination characteristic for iron-55.

THE COUNTING EFFICIENCIES

The list, with results, of counting efficiencies is given in Table 1. All measurements are for PBD in toluene scintillator with 1 ml of pure alcohol.

To obtain the counting efficiency the standard solutions, of the radionuclides in question, were used with radioactive concentration determined by the $4\pi(x,e)\text{-}\gamma$ coincidence method or 4π plastic scintillation counter. As it is shown in Table 1 the counting efficiency increased with the energy of radiation and seemed to be similar for K-X photons as for Auger-electron counting. To prove this observation the additional experiment was done, where only K-X of iron-55 photons were counted. The iron-55 source was prepared on transparent plastic foil, covered with another foil and placed into liquid scintillator. Auger-electrons were fully absorbed in the foils and only K-X photons were counted. Taking K-X yield ($\frac{\text{K-X}}{N_0}$) into account the obtained counting efficiency was very close to that shown in Table 1.

THE QUENCHING EFFECT

To prove the possibility of application of the liquid scintillation method for activity measurements of E.C. nuclides, the quenching effect has been checked for the nuclides in questions. This was done by counting the standard samples with additional amounts of carrier of the same chemical composition. The quenching effect for manganese-54 is shown in Fig. 3.

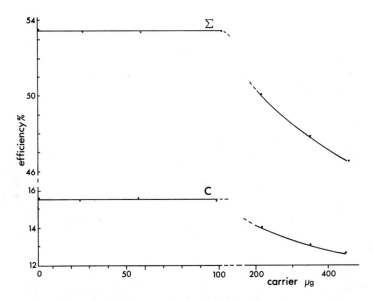

Fig. 3: The counting efficiency for manganese-54 and PBD in toluene as a function of the Manganese carrier. Σ: The parallel system. C: The coincidence system.

The counting efficiency was stable and the quenching effect was not observed up to 100 μg of carrier. It should be stated that the amount of carrier in the normal counting sample is not more than several μg. Table 1 became an additional test for the quenching effect, as the results obtained did not depend upon the mass of the samples. It should be

Table 1. The counting efficiency with PBD in toluene as scintillator.

No of sample	Nuclide	Energy K-X (keV)	Energy (keV)	Amount of samples	Mass of samples (g)	Chemical composition	Counting efficiency coinc. %	Counting efficiency paral. %	Supplier of stand. solut.	Error of stand. %	Counting error coinc. %	Counting error paral. %
1	chromium-51	4.95	325	4	0.01518 →0.02543	25 µg Cr as K_2CrO_4 + 25 µg Cr as $CrCl_3$ in 1 g 0.1 NHCl + 40 µg NaCl	14.9	—	R.C. Amersham	±1.4	±1.2	—
2	manganese-54	5.4	842	4	0.00640 →0.02595	25 µg Mn as $Mn/NO_3/_2$ in 1 g 0.01 $NHNO_3$ + 0.1% formaline	15.5	53.6	R.C. Amersham	±2.8	±1.3	±2.0
3	iron-55	5.9	—	6	0.01470 →0.02656	25 µg Fe as $FeCl_3$ in 0.1 N HNO_3	16.0	56.8	IBJ Warsaw	±1.8	±1.4	±2.5
4	zinc-65	8.05	1115	4	0.00590 →0.03195	20 µg Zn as $ZnCl_2$ in 1 g 0.1 N HCl	31.7	60.5	IBJ Warsaw	±1.1	±2.2	±1.6
5	yttrium-88	14.2	1836 898	5	0.013799 →0.050588	10 µg Y in 1 g 0.1 N HCl + 0.1% formaldehyde	39.6	—	IAEA Vienna	±2.0	±1.7	—
6	iodine-125	27.4	35.3	4	0.00536 →0.02099	25 µg I as KI + 50 µg $Na_2S_2O_3$ in 1 g solution + 0.1% formaline	50.3	73.3	R.C. Amersham	±2.7	±2.5	±3.5

also emphasised that the quenching effect was stronger for the coincidence system, than for the parallel one, as is shown by the counting ratio of the two counting systems in Fig. 4.

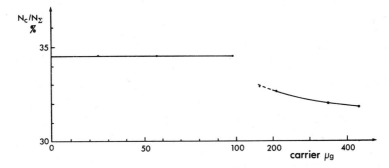

Fig. 4: The ratio of the coincidence system counting rate to the parallel system counting rate for manganese-54 (PBD in toluene scintillator) as a function of the carrier.

CONCLUSIONS

As it was pointed out, the liquid scintillation counting is a convenient method for activity measurements of E.C. nuclides. Although full counting efficiency was not obtained for any investigated nuclide, and the absolute method could not be realised, the obtained counting efficiencies were probably the highest from any possible counting method. The obtained counting efficiency for iron-55 can be compared with that for the scintillation counter with plastic scintillators,[2] where the counting sample was prepared in plastic foil and covered with another foil. The overall efficiency was lower at the case of plastic scintillators, as Auger electrons were completely absorbed in the foils, but full counting efficiency was obtained for K-X rays. This discrepancy was probably due to the light collection ability, which was much more effective for a point source and plastic scintillator sandwich, than for a distributed source at the liquid scintillator with an additional energy loss in wall effects.

An additional advantage of the 4π liquid scintillation method in the case of iron-55 is the possibility of activity measurement, independently from the yield factor (W_k) which is taken as a basis for activity calculation in the methods actually in use. The values for the yield factor (W_k) published by several authors show discrepancies of up to 10%. Providing that the counting efficiency for K-X photons and Auger-electrons are very close to each other, the counting efficiency for iron-55 can be obtained by interpolation of the values for other nuclides (chromium-51, manganese-54, cobalt-57) which can be standardised by the $4\pi(x,e)$-γ coincidence method.

It should be also noticed that application of Triton X-100 was not successful in one investigation. The counting efficiency was generally lower (about 50% that of toluene scintillator) and the quenching effect investigation did not give reproducible results.

REFERENCES
1 R. A. Allen, *Metrology of Radionuclides,* International Atomic Energy Agency, Vienna, (1960) p. 343.

2   A. Tada and T. Radoszewski, *Standardisation of Radionuclides*, International Atomic Energy Agency, Vienna, (1967) p. 293.
3   T. Radoszewski, Chapter 2 of this book.
4   T. Radoszewski, *Nukleonika* **5**, 361 (1960).

DISCUSSION

E. Langenschiedt: Which effects in your scintillation methods do you call 'wall effects'? Can you define them? Do they influence your extrapolation of the discriminator curve.

T. Radoszewski: The wall effect in standardisation is a geometrical concept and it is due to the loss of particles or photon energy in the wall instead of the scintillator. It could not be covered by the extrapolation curve—or, at least, only partly.

J. A. B. Gibson: In addition to wall effects there is a reduced efficiency at low energies due to ionization quenching when the rate of energy deposition (dE/dX) is high. The specific light output (dL/dX) is given by: $dL/dX = S \cdot dE/dX/(1+kB\, dE/dX)$ where S is the scintillation efficiency and kB is a constant. The effect is shown in Fig. 6 in the discussion to Chapter 2.

T. Radoszewski: Yes, but still for α-emitters the wall effect would not have any influence on counting efficiency.

W. R. Greig: Is it possible to determine the counts due to iodine-125 and to tritium respectively when both are present in solution in a toluene PPO/POPOP system?

T. Radoszewski: I have not investigated this problem, but owing to the very poor resolution power for small energy in liquid scintillators, I think it will be difficult to distinguish the pulses from iodine-125 and tritium.

B. W. Fox: I can comment that to count iodine-125/tritium mixtures, the iodine-125 standard is counted in a gamma-counter and also in the tritium channel of a liquid scintillation counter. The tritium counts calculated from the instrument rates are subtracted from the total counts in the tritium channel to assess the tritium present. Also with Triton X-100 systems, it is the figure of merit which matters most, not absolute efficiency. What figure of merit can be achieved in this system?

T. Radoszewski: I was not interested in the figure of merit as I usually count very small amounts of solution (about 10 mg). I just measured to check the efficiency of counting for Triton X-100.

J. A. B. Gibson: Referring to iodine-125 and tritium counting, the resolution for the iodine-125 (i.e. X-ray at 27.4 keV) would be approximately 50% and thus for small amounts of tritium there will be a large contribution from iodine-125 in the tritium channel.

M. I. Krichevsky: In double label counting of isotopes with overlapping spectra it is best to allow maximum overlap in the various channels (without identity) rather than try to achieve maximum separation. Simultaneous equations are used to separate the counts due to each isotope in each channel and then compensated for quenching.

Chapter 7

Optimisation Techniques for Computer-Aided Quench Correction

Micah I. Krichevsky* and Charles J. MacLean†

Environmental Mechanisms Section, Office of the Director of Intramural Research
and Population Genetics Section, Human Genetics Branch,†
National Institute of Dental Research, National Institutes of Health,
Bethesda, Maryland 20014 USA

In liquid scintillation counting as practiced in today's laboratories the reliability and accuracy of the machines is commonly assumed to be a negligible factor when compared with the accuracy of the measurement being performed. Fortunately, for the majority of cases this is correct. The user of scintillation counters is allowed the luxury of some powerful simplifications. He can decide how long to count any specific sample based on Poisson statistics. He ignores the machine drift, sample asymmetries, position errors, sample inhomogeneities and most of the other pitfalls in the path of righteous counting. By controlling sample composition he will even often be able to ignore variable quenching.

All of these problems, however, become manifest for computer programs with the ultimate goal of on-line data capture and reduction. The most important design requirement for such a package is that the programs be written for the most sophisticated user to be encountered. With today's high speed computers, the storage space and management which would be required to hold a series of programs of various levels of sophistication are much more expensive than would be the saving in computer time realised by most of the users being content with simple algorithms.

Thus, it becomes incumbent on the systems analyst to find the limits of the system and those factors contributing to error. Using such a systems approach leads, in turn, to optimisation of the data handling, the machine design, and finally (and most difficult in many cases) education of the user in proper machine utilisation and sample preparation.

The goal is a good estimator of the disintegration rate D. This rate is observed indirectly by counting and adjusting for the efficiency, $D = C/E$ where C is the counting rate, E the efficiency of the channel. Goodness is not a subtle concept in this case. A good estimator should be sufficient, be consistent, have minimum variance, and perhaps be unbiased. We want to know its sampling distribution, or at least its variance so as to set sample sizes and confidence limits and devise hypothesis tests. A standard procedure for deriving a good estimator is the maximum likelihood method. Maximum likelihood estimators are commonly used since they inherently have the desirable qualities listed above, with the possible exception of bias in the particular case. This method is based upon the concept of the likelihood, which is a function relating the set of observations to the parameter to be estimated. Maximum likelihood estimator means that value of the parameter for which this function achieves its maximum.

Automatic external standard will be treated first and channels ratio quench correction later. Other forms of quench correction, e.g., internal standardisation, do not lend themselves as readily to on-line computation and will not be considered here.

The sample efficiency is a known function (e.g. a cubic is commonly fitted) of the true AES (i.e. the count rate of one channel or ratio of two channels of automatic external standard) which is estimated by observing a variable, S, AES counts or ratio. We know that the average of observations on S is the maximum likelihood estimator of the true AES and therefore the sample efficiency at this average, $E\,[\text{ave}(S)]$, is the maximum likelihood estimator of E.

To show this, let L represent the likelihood and recall that the maximum of a function is achieved where its derivative vanishes. Because of the chain rule of calculus,

$$\frac{\partial L(E(\text{AES}))}{\partial \text{AES}} = \frac{\partial L(E(\text{AES}))}{\partial E} \cdot \frac{\partial E(\text{AES})}{\partial \text{AES}},$$

we see that

$$\frac{\partial L}{\partial \text{AES}} = 0 \text{ where } \frac{\partial L}{\partial E} = 0.$$

Secondly, C is estimated by observing the variable X, the number of sample counts. The maximum likelihood estimator of C is the average, $\text{ave}(X)$. Our goal, however, is neither C nor AES, but rather $D = C/E(\text{AES})$ where C and AES are nuisance parameters to the estimation problem.

Although X and S are statistically independent, we must solve for both simultaneously. The distribution of X is a function of the unknown AES which is estimated by S. Consider the likelihood function: $L(X, S \text{ given } D \text{ and AES})$.

Because S is not a function of D we have:

$L = L(S \text{ given AES})\, L(X \text{ given } D \text{ and AES})$.

$L(X \text{ given } D \text{ and AES})$ is simply the Poisson distribution with parameter D times AES. $L(S)$ is not a common form, however recall that we know that the maximum likelihood estimator for AES is $\text{ave}(S)$, the average of the observations S. Therefore by the same argument as above we also have

$$\frac{\partial L\,(S \text{ given AES})}{\partial \text{AES}} = 0 \text{ at AES} = \text{ave}(S)$$

Moreover:

$$\frac{\partial L\,(X \text{ and } S \text{ given } D \text{ and AES})}{\partial D} \quad \text{is proportional to}$$

$$\frac{\partial L\,(X \text{ given } D \text{ and AES})}{\partial D}$$

so that the simultaneous solution will have

$$\frac{\partial L\,(X \text{ given } D \text{ and ave}(S))}{\partial D} = 0$$

This is easily seen to be satisfied by

$$G = \frac{\text{ave}\,(X)}{E\,(\text{ave}\,(S))}.$$

We therefore use G as the maximum likelihood estimator for D.

Since X and S are independent variables, we can derive a good first order estimate of the variance of the maximum likelihood estimator of D as follows:

$$\text{var}\,(G) = \left[\frac{\partial D}{\partial E} \cdot \frac{\partial E}{\partial S}\right]^2 \cdot \text{var}\,(S) + \left[\frac{\partial D}{\partial C}\right]^2 \cdot \text{var}\,(X)$$

$$= \left[\frac{\text{ave}\,(X)}{E\,(\text{ave}\,(S))}\right]^2 \cdot \left[\frac{\partial E}{\partial S}\right]^2 \cdot \text{var}\,(S) + \frac{\text{var}\,(X)}{E\,(\text{ave}\,(S))}$$

Since X is a Poisson variable its variance equals its mean. We can therefore estimate the variance with the same mean as estimates ave (X). If we also use a Poisson count as the AES reading, then var (S) = ave (S) and the form reduces to

$$\text{var}\,(G) = \left[\frac{\text{ave}\,(X)}{E\,(\text{ave}\,(S))}\right]^2 \cdot \left[\frac{\partial E/\partial S \text{ ave}\,(S)}{E^2\,(\text{ave}\,(S))} + \frac{1}{\text{ave}\,(X)}\right]$$

When a ratio is used for AES, this makes the calculation of its variance more difficult, and the easiest solution is just to estimate var (S) from the sample of S.

In channels ratio counting, the efficiency estimator and the sample count rate estimator are not independent since the efficiency estimator (the observed ratio) is derived from the observed sample count rate estimator plus the count rate in a second channel. This makes the derivation of maximum likelihood estimator of the disintegration rate more complicated but still quite analogous to that described above for AES quench correction. The channels ratio counting maximum likelihood estimator, V, is:

$$V = \frac{1}{n} \sum_{i=1}^{n} X_i / E\left(\sum_{j=1}^{n} W_j / \sum_{k=1}^{n} X_k\right)$$

In this case W and X are naturally not independent, the parameter estimated is

$D = C_x/E\left(\dfrac{C_w}{C_x}\right)$, and X is the larger channel and W the smaller.

The variance of the channels ratio estimate is calculated essentially in the same way as for the AES method, except the counts and efficiency estimates are not independent. Therefore, their covariance must be included.

Given the form:

$$S = \dfrac{\text{ave }(X)}{E\left(\text{ave }(W)\,/\,\text{ave }(X)\right)},$$

we approximate the variance:

$$\text{var }(S) = \left[\dfrac{\partial S}{\partial C_x}\right]^2 \cdot \text{var }(X) + \left[\dfrac{\partial S}{\partial C_w}\right]^2 \cdot \text{var }(W)$$

$$+ \left[\dfrac{\partial S}{\partial C_x}\right] \cdot \left[\dfrac{\partial S}{\partial C_w}\right] \cdot 2\,\text{cov }(X,\,W)$$

To evaluate this, we see that dS/dC is estimated by:

$$\dfrac{\partial S}{\partial \text{ ave }(X)} = \dfrac{E\left(\text{ave }(W)\,/\,\text{ave }(X)\right) - \text{ave }(X) \cdot \left[\dfrac{\partial E}{\partial \text{ ave }(X)}\right] \cdot \left[-\text{ave }(W)\,/\,\text{ave}^2(X)\right]}{E^2\left(\text{ave }(W)\,/\,\text{ave }(X)\right)}$$

$$= \dfrac{E\left(\text{ave }(W)\,/\,\text{ave }(X)\right) + \left[\dfrac{\partial E}{\partial \text{ ave }(X)}\right] \cdot \left[\dfrac{\text{ave }(W)}{\text{ave }(X)}\right]}{E^2\left(\text{ave }(W)\,/\,\text{ave }(X)\right)}$$

and similarly:

$$\dfrac{\partial S}{\partial \text{ ave }(W)} = -\dfrac{\partial E / \partial \text{ ave }(W)}{E^2\left(\text{ave }(W)\,/\,\text{ave }(X)\right)}$$

Therefore we have:

$$\text{var }(S) = \dfrac{1}{E^4} \cdot \left[\left[E + \dfrac{\partial E}{\partial \text{ ave }(X)} \cdot \dfrac{\text{ave }(W)}{\text{ave }(X)}\right]^2 \cdot \text{var }(X) + \left[\dfrac{\partial E}{\partial \text{ ave }(W)}\right]^2 \cdot \text{var }(W)\right.$$

$$\left. + 2\left[E + \dfrac{\partial E}{\partial \text{ ave }(X)} \cdot \left[\dfrac{W}{X}\right]\right] \cdot \left[\dfrac{\partial E}{\partial W}\right] \cdot \text{cov }(X,\,W)\right]$$

In the Poisson process, the variance is equal to the mean. Therefore, we can estimate

var (X) = ave (X)

and

var (W) = ave (W)

Moreover, the covariance between Poisson counts when one window is entirely within the other is the mean of the smaller.

cov (X, W) = ave (W)

So that, finally:

$$\text{var}(S) = \frac{1}{E^4} \cdot \left[\left[E + \frac{\partial E}{\partial \text{ave}(X)} \cdot \left[\frac{\text{ave}(W)}{\text{ave}(X)} \right] \right]^2 \cdot \text{ave}(X) \right.$$

$$\left. + \left[\frac{\partial E}{\partial \text{ave}(W)} + 2E + \frac{\partial E}{\partial \text{ave}(X)} \cdot \left[\frac{\text{ave}(W)}{\text{ave}(X)} \right] \right] \cdot \left[\frac{\partial E}{\partial \text{ave}(W)} \right] \cdot \text{ave}(W) \right]$$

The derivatives of E with respect to the windows depends on the functional form. It can be easily evaluated in any particular case.

Propagation of error through calculations. In the discussion to follow all arguments apply both to AES and channels ratio quench correction.

The maximum likelihood is one of the statistics currently used, however it is not the only common one.

Perhaps the most subtle of all error sources (except for the statistician or numerical analyst) to cope with is the propagation of error by the calculations themselves. An example (1) will illustrate the proposition.

Example 1   Let $y = [(x-9) \times 10] - 9$
Then if $x = 10$ and is error free
$y = 1.0$
However, if x has an error of -0.1 (i.e., -1%) then
$x = 9.9$ and
$y = 0$. Thus, this simple series of mathematical manipulations has magnified the error 10 fold and reduced the calculation to nonsense.

In order to apply the latter considerations to liquid scintillation counting we began by asking a deceptively simple question. If one assumed a perfect counter with no drift or other sources of error, how should one calculate the results of a series of repeat counts? The existence of a polynomial (e.g. a cubic equation) which describes the relationship between absolute efficiency and the AES is also assumed. That is, should one total the observed counts and divide by the total time counted, thus calculating d.p.m. once only? Or is it equivalent to calculate the d.p.m. for each time the sample is counted and average the calculated d.p.m. figures?

Stated another way, if a radioactive sample is counted a number of times, there are at least three ways a computer program may be written to perform the estimation of the disintegration rate. First, in the maximum likelihood estimation described above, the observed counts are accumulated over the total observations and divided by the total time counted and the final calculation performed only once. Second, the disintegration rates could be calculated for each observation and these averaged at the end. Third and intermediate between the other two, is the possibility of averaging the count rates and averaging the efficiencies separately and then dividing. The difference in computer time for the alternatives is negligible, but there are other important differences.

One of the important considerations of a statistic is the limiting value it has when the sample consists a very large number of observations. The idea is that you should get the 'true' value if you sample enough. It happens that the results are not the same if the order of operations is reversed. All maximum likelihood estimations, ours included, converges on the true values. Consider the second estimator for D

$$Y = \frac{1}{n} \sum_{i=1}^{n} \frac{X_i}{E(S_i)}$$

where as above n is the number of observations, E is the efficiency, S is the observed AES and X is the sample count. As n is increased the value of Y will be distributed more and more tightly around the mean

$$Y \rightarrow \text{ave} \left[\frac{X}{E(S)} \right].$$

Since X and S are independent, the mean of their ratio is the ratio of their means

$$Y \rightarrow \text{ave}(X)\, \text{ave} \left[\frac{1}{E(S)} \right].$$

However the second term, the mean of the inverse of the efficiency, is not the value we need which is the inverse of the mean efficiency. Therefore, as sample size is increased the value Y will differ from D more and more certainly. Thus, the more the sample is counted and calculated by this method the surer it is that the answer converged upon will be wrong. The third common alternative estimator (T) above is

$$T = \sum_{i=1}^{n} X_i / \sum_{j=1}^{n} E(S_j).$$

The same problem occurs again because the limit of the denominator is the mean of efficiency and not the efficiency of the mean of S, which is what we want.

Sources of error and minimisation of their effect. Table 1 presents major categories of input to the scintillation counting system as contributing error and some means of minimising the error due to each of the categories. Errors inherent in the sample itself can only be minimised by careful sample preparation (including container selection). The one exception to this is the detection and recovery of error due to inhomogeneous samples, as will be discussed later.

Three classes of machine error exist: *(a)* additive (e.g. background); *(b)* subtractive (e.g. coincidence loss); and *(c)* spectral distortion resulting a drifting response to pulse height (e.g. drift in photomultiplier gain). Additive errors such as 'background' are normally minimised by the approximation of determining its magnitude in each counting

Table 1. Methods of recovery from error in liquid scintillation counting.

Sample
    Preparation   Accurate, multiple measurements
                                        Homogeneous distribution in vial
    Container   Accurate manufacture
                                        Selection by user
    Position   Recounting after rotation of vial
Machine
    Electronics   Preventive maintenance
                                        Stable design
    Photomultipliers   Cooling
                                        Periodic drift correction
                                        Short counts, multiple cycles
                                        Intersperse standards
Data reduction
    Calibration for quenching
        AES   Many standard vials
        Channels ratio   High count rates in each channel
    Calculations   Test method for minimal error propagation

---

channel on an 'unquenched' sample and subtracting these figures from all other determinations. This method ignores the fact that 'background' pulses have two main sources, i.e. extraneous radiation in the vial and machine generated pulses. The approximation is valid only for the machine generated pulses. Extraneous pulses due to radioactive events must be treated by a quench correction algorithm since their magnitude changes with quenching in a manner analogous to the isotope(s) being determined.

Subtractive loss, most commonly coincidence loss, are simply avoided by keeping the sample count rate below levels at which the losses become discernible. Due to the high speed electronics in modern counters this imposes essentially no restriction and hence requires no further treatment in this discussion.

Drift in response is the most serious of the machine generated errors. It is most commonly due to changing response in the photomultiplier tubes. Thus, it leads to 'unique' errors which depend on the nature of the sample and the window settings. Recovery from this type of error requires a combination of proper counting procedures and a computational method designed to minimise error propagation.

The most obvious source of increased error in data reduction is the use of inaccurate methods for calibration curve fitting. For example, Wampler[2] has evaluated a great variety of polynomial least squares methods and found enormous variation in their results.

The theoretical distribution of the variables involved in scintillation counting are quite simple. Exponential decay generates data distributed according to the Poisson Law. When two channels are used the smaller channel usually is completely inside of the larger channel which is therefore made up of counts in two channels which are independent. When AES is used it is independent of the number of counts in the channels.

(Because the channels ratio estimator is derived directly from the sample counts it is affected by inhomogeneity in the sample. The AES estimator, being independent, is largely unaffected by non uniform samples. Thus major discrepancies in calculated disintegration rates using the two efficiency estimators would indicate biphasic or inhomogeneous samples. Provided the count rates are high enough to give satisfactory estimates of quenching, the channels ratio results should be used and will largely be valid).

Now all the distributions depend heavily upon these assumptions. All the expectations and particularly the covariances would be affected by deviations of the data from these theoretical qualities. Many tests of repeated counting were performed over long periods and it was easy to trace the deviations from the values expected due to very large drift of the counts over time. Although there are several hypotheses for explanation of this drift, no one has proved the cause definitely.

Table 2 is a summary of the correlation analysis performed on the output of two experiments which were originally run without concern for time trends. They were run rather as a verification of the theoretical distributions with no expectation of what actually occurred. They were for that reason, not truly experiments in the sense that they were designed to optimise any sort of an observation. Several vials with different quenchings and different radioactivities were just mixed together in the automatic machine which ran them interleaved. They ran in order 1 to 6 and 1 to 4 respectively.

The vials $0-1$ and $N-1$ listed in Table 2 had no radioactivity in them at all and so the correlations are not very meaningful. The 24 counts and 17 counts in the wide channels are simply background.

In all the other runs the correlations and variances are quite outside of the values which would have been obtained if the theoretical model were in fact operating. First of all the theoretical correlation between the AES and the sample counts is zero. They are independent events, being quite different measurements of exponential decay in completely different windows and at different times. There should be no correlation at all. Although the correlations obtained are not consistent they are often quite significant. The AES, in this case, is a ratio between two channels in the range induced by gamma radiation. The correlation between AES and the channels ratio of the wide and narrow windows in the beta range is often around 40 to 50%. This could lead to an explanation that the same phenomenon is taking place all along the energy spectrum throughout time. For example, if the overall gain of the photomultipliers were changing over a period of time, and if the two pairs of windows of the AES and the channels had the same relationship to each other, then a downward shift in both cases would have a correlated effect over a period of time. Another problem is the correlation between the wide and narrow channels. The correlation between the two Poisson variables, one inside of the other, as is the case here (where the narrow channel is a part of the wide channel) should be the square root of the ratio of the two channel means. Taking the average for each channel as a estimate of the mean, the correlations actually obtained are much too low in every case. In addition, the variance of a Poisson variable is equal to the mean of that variable. Taking the averages and standard deviations of the values obtained, the standard deviation of the channels are much too high for them possibly to have come from a Poisson distribution.

The experiment number 36, the last one in Table 2, is a test of the first hypothesis brought forth to explain the correlations observed. The hypothesis was that the move-

Table 2. Statistical analysis of repetitive counting.

A series of vials used as calibration standards for carbon-14 were counted repetitively for approximately 2.5 days. The vials were counted for 1 min on each cycle. Thus, each vial was counted about once every 12 minutes. All data was taken from Packard 3375 Liquid Scintillation counters. (Comparable data however, was obtained using counters of other manufacture. Since it is not the purpose of this chapter to evaluate various brands or types of counters, comparison will be specifically avoided.)

The columns denoted: (0–n); (N–n); (N–1) and 36 were each a separate series of determinations. All vials contained the same amount of isotope, except the non-radioactive vials 0–1 and N–1. The data are rounded to 3 significant figures in most cases.

Experiment	0–1	0–2	0–3	0–6	N–1	N–2	N–3	N–4	36
Average									
Wide (c.p.m.)	24.6	32,400	32,600	27,800	17.2	32,600	28,600	15,600	31,000
Narrow (c.p.m.)	20	29,100	28,300	16,600	8.51	17,100	10,000	968	23,000
AES (ratio)	0.65	0.62	0.60	0.37	0.62	0.61	0.40	0.30	0.63
Ratio (n/w)	0.81	0.89	0.87	0.60	0.49	0.52	0.35	0.062	0.75
Standard Deviation									
Wide	5.22	179	174	193	4.26	189	160	151	192
Narrow	4.43	182	183	195	2.95	156	123	37	189
AES	0.0022	0.0022	0.0021	0.0021	0.0021	0.0023	0.0018	0.0018	0.0080
Ratio	0.26	0.0044	0.0044	0.0045	0.12	0.0037	0.0034	0.0022	0.0031
Correlation									
AES, Wide	+ 0.05	– .09	– 0.15	+ 0.37	0.00	+ 0.21	+ 0.22	+ 0.44	– 0.29
AES, Narrow	+ 0.04	+ .27	+ 0.28	+ 0.58	0.00	+ 0.49	+ 0.43	+ 0.43	+ 0.16
AES, Ratio	0.00	+ .44	+ 0.52	+ 0.56	+ 0.03	+ 0.46	+ 0.43	+ 0.34	+ 0.12
Wide, Narrow	+ 0.54	+ .66	+ 0.64	+ 0.79	+ 0.71	+ 0.63	+ 0.63	+ 0.43	+ 0.87
Wide, Ratio	– 0.50	– .29	– .24	+ 0.32	0.00	0.00	+ 0.22	+ 0.20	+ 0.19
Narrow, Ratio	+ 0.43	+ .53	+ .77	+ 0.83	+ 0.66	+ 0.67	+ 0.89	+ 0.96	+ 0.66

ment of the vial and the positioning of the vial on each cycle induced the correlation. It was hypothesised that the measurement would be different depending upon the exact position or circular angle at which the bottle was placed in the receptacle for measurement. A test was made over about the same period of time in which the vial was not moved at all. The errors and correlations are approximately the same for that experiment, with the exception of the standard deviation of AES in which a mechanical failure was found. The data are still not in Poisson distribution and a plot of the data shows that the time trend is still in action at what would appear to be at the same level. The movement of the vials seems to have explained very little.

Plotting the data in a serial fashion led to a very clear picture of what was disturbing the theoretical distribution. The powerful time trend completely overwhelms the random deviation of the data in both cases being of a completely larger scale over the whole duration of the experiment. This can easily be seen in Fig. 1.

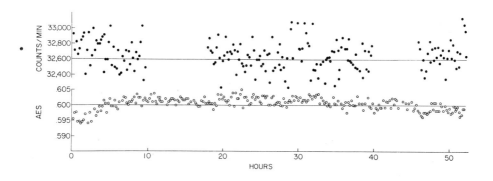

Fig. 1: Long Term Time Trends in Counting. This data is a plot of some of the actual observations from vial 0–3 in Table 2. The AES numbers are the ratios of two channels. The horizontal lines are the averages of the data. The gaps in the data have no significance other than the level of persistence of the authors in plotting data.

This calls immediately for a new model to explain the variations of the data observed. The simplest model to appeal to one is the following:
$Y = P + T$ where Y is the observation, the component P is distributed as a random Poisson random variable with parameter C, plus a component T which is the drift random variable. So far, we can put no conditions on the drift factor except that its mean value equals zero. By calculating the time trend and subtracting it out of the model (or subtracting an estimator of it out) this should leave the Poisson variable we had originally believed that we had. This procedure would be followed for the AES estimator and for every applicable sample channel. The exact nature of the time trend will vary with the particular variable being estimated.

Naturally the important question is how to obtain an estimate of the drift. A common method would be a running average of the data. For time trend component of observation j, take the average of all the values from $j-10$ to $j+10$ to be an estimator of the trend at time T_j. A standard error for that estimate can easily be obtained.

The time trend will be removed under the conditions that the expected value of X at any time T_1 equals the expected value of X at any other time T_2, that those expected values over ranges of T must all be equal and finally that the autocorrelation between times with lag T must be zero for all time lags.

The foregoing discussion assumes that the half-life of the isotope is great enough to be a negligible part of the drift of the system. Counting of short half-life isotopes adds a complication to the search for better estimators of decay rate. On first consideration, it appears to us that the data should be corrected back to some arbitrary but common starting point in time as early as possible in the algorithm. This minimises weighting the answer obtained by the time at which the particular sample is counted. In counting of one isotope at a time, the correction back to 'zero time' thus should be done on the observed counts themselves.

In multiple isotope counting of short half-life isotopes, the half-life correction cannot be made on the raw counts since the ratio of counts due to each isotope in each window is unknown until the estimator of quenching for each observation is evaluated. The ratio of isotopes at the time of counting would then be evaluated for each sample and the counts due to each corrected back to the initial point before final calculation.

Evaluation of alternate algorithms for computing disintegration rates is made difficult due to the lack of a precise radiation standard. The most accurate standards currently available are stated to be ± 0.5% with respect to disintegration rate. A further complication is introduced by the unknown error in sample preparation. Drift due to chemical instability versus machine is undefined.

It would be of advantage to develop sets of data simulating the various types of error that might be encountered in practice. Thus, the exact correct answer would be known. Various algorithms could then be tested empirically and directly for their ability to handle error and converge on the correct value.

To this end, we have built a simulator program which develops the requisite data. It allows up to twenty channels to be simulated with any degree of correlation.

The potential advantage of simulation becomes evident when one considers the complexity of the range of questions to be asked of a simulator. As mentioned above, various algorithms can be compared for their ability to converge on a correct solution. The theoretical derivation of a good estimator for each possible algorithm would not be a practical approach. Other possible algorithms being considered are interpolation after spline fitting or some form of linear interpolation using polygonal line segments.

The simulator generates counts similar to real data in that it is Poisson distributed, it may vary with time, and the various windows can be either independent or related. The value of such generated data over real data, in addition to convenience, is that the estimation procedures can be related to the true values and their performance thereby judged. With real data, when one estimates, there is no true value with which to compare.

In this simulator, the time trend is defined by the user through its expression by Fourier series coefficients. There are available standard computer programs which calculate these coefficients for any set of sample data, so that the user need only supply data having the characteristics he would like to test. It can be taken from real observed drift if appropriate. The entire energy spectrum being tested must be defined by the user, but only in terms of the desired covariance relationships among the windows used need be specified and this can be taken from observations. With the true decay rate calculated for each period, an 'experiment' is run by using random numbers to generate a count which deviates from the expected value by a random amount. A set of these values are then subjected to the data analysing program to be evaluated by comparing the estimated values with the true input values. This method is referred to as a numerical experiment.

Another aspect of the use of simulation, in addition to comparing alternative algorithms, is the ability to assess the contribution to final error of various sources of error. It would be most difficult to calculate the contribution of drift, for example, in a given counting window to the overall error in multiple isotope counting. With cubic polynomials fitted to the efficiency function for each isotope in each counting window,[3] the two isotope case leads to evaluation of a sixth degree polynomial.

Some interim conclusions can be reached at this writing. Because of machine drift and half-life considerations, counting of high precision should not be attempted using any method in which the decay rate is observed directly. This admonition holds whether the decay rate is derived from a special purpose computer installed on the counter itself, by an on-line general-purpose computer, or any other method of computation.

Throughout, we have assumed that, at any given level of quenching, all similar samples in a given counting window will exhibit similar drift characteristics. While the assumption has not been tested *per se*, it is related to the principle underlying quench correction by both AES and channels ratio. Recovery from drift errors is possible by including quench standards with the unknowns. The whole series is counted for a short time on each sample. The whole procedure is repeated for as many observations on the series as is required to accumulate the total time period desired. The raw data is averaged and the quench correction done but once. The curve fitting should then correct for drift as well as quenching.

REFERENCES

1   Richard L. Schrager, Division of Computer Research and Technology, National Institutes of Health, Bethesda, Maryland, Private communication.
2   R. H. Wampler, *J. Res. Nat. Bur. Std. B* **73B**, 59 (1969).
3   M. I. Krichevsky, S. A. Zaveler and J. Bulkeley, *Anal. Biochem.* **22**, 442 (1968).

DISCUSSION

B. Scales: Are you suggesting that a single 10 min count is inherently more inaccurate than the sum of 10 x 1 min counts for simple non-computerised work, especially when samples are being analysed which are close to background?

M. I. Krichevsky: Yes. In a drifting system one must calibrate and count at the 'same' time. Additionally, if one is close to background, it is essential to calibrate quenching of the background itself.

G. G. J. Boswell: Is your computer for the exclusive use of the five liquid scintillation counters?

M. I. Krichevsky: No, amino acid analysers, neurophysiology laboratories, a gas chromatograph, a u.v.-visible recording spectrophotometer and some fermentation equipment are in various stages of being interfaced with our process-control computer. Five counters on one computer is a trivial use of any computer having the capacity to handle the computations on-line because of the extremely slow data rate.

G. A. Buckley: Could you put the right perspective by telling us the error when one has chosen the method of calculation which 'converges on the wrong answer'?

M. I. Krichevsky: The error is obviously small (usually less than a few percent for singly labelled counting). By simulation we hope to give a more intelligent answer to this question in the future.

D. S. Glass: (i) How accurate do you consider you can make your liquid scintillation counter operate—as an intrinsic property of the machine and the calibration method? (ii) Do you consider a twenty channel (hypothetically) machine would overcome drift problems by comparing results in each channel? (iii) How many liquid scintillation counters do you need to justify an on-line system (as a question of cost)?

M. I. Krichevsky: (i) Certainly better than ± 1% if all pains are taken. How much better I cannot say; in fact, it becomes a matter of definition since the best standards are not as good. Thus we can only talk in relative terms. (ii) No. Machine drift is inseparable from spectral shift due to quenching or chemical instability if one considers only the sample data. Only recalibration (either mechanically or by computation) against a stable standard can do this. (iii) If the central computer exists and will accept teletype or equivalent input in its operating system then 2 to 4 should do as most interfaces for computers to accept such data can handle multiple 10 to 15 characters per second data rates. Such an interface can be obtained for less than $10,000. If the computer is doing nothing else but processing this kind of data, then one would presumably need 10 or more to justify the expense. The realistic answer is conditioned however, more by the particular administrative environment one is in and one's station within it—coupled to the persistence of the request rather than the number of counters.

K. L. Evans: Could you provide information on the selection of polynomials as models for quench correction curves, i.e. the basis upon which the polynomial order is selected?

M. I. Krichevsky: If polynomials are used for calibration, the order should be increased until the standard error is within acceptable limits. This assumes that the data is well-behaved. I would suggest that most quench curves that have been obtained can be fitted to a cubic with a degree of accuracy that makes it non-limiting in terms of overall error.

Chapter 8

The Processing of Liquid Scintillation Spectrometer Data Using a Desk-Top Computing System

M. A. Williams and G. H. Cope

*Department of Human Biology and Anatomy,
University of Sheffield, England.*

INTRODUCTION

Liquid scintillation spectrometry is now widely used in biochemical and chemical laboratories for assaying soft β-emitters. Most modern spectrometers are sophisticated instruments utilising several channels and capable of automatically handling hundreds of samples and printing out the results. These instruments usually incorporate a means of estimating the degree of quenching, the most modern by means of the 'external standard-channels ratio' method. Some expensive new machines incorporate computation facilities that both determine and allow for the degree of quenching and can also relate the results to other variables such as sample weight or specific radioactivity. Some spectrometers can be linked directly, or via punched tape, to a large computer whilst others can be linked to desk-top computing systems. These latter systems have not to date proved entirely satisfactory. Limited programming steps and storage space has prevented both accurate expression of quench curves and the analysis of binary-labelled samples.

The work reported here is directed towards those workers with neither the money to purchase expensive machines with built-in computing facilities nor with regular access to a large computer. The system we have evolved uses a moderately priced desk-top system. Its use permits the owner of the more reasonably priced scintillation counter to have effective computing facilities for singly and doubly-labelled samples. Additionally he gains a desk-top computer for general laboratory work.

METHODS

Labelled materials. 1,2-^{3}H-n-hexadecane (specific radioactivity 2.45 μ Ci/g) and 1-^{14}C-n-hexadecane (specific radioactivity 0.97 μCi/g) (The Radiochemical Centre, Amersham, Bucks, U.K.) were used as standards.

The livers from rats that had been injected intraperitoneally with Me-^{3}H choline chloride and 2-^{14}C-ethan-1-ol-2-amine hydrochloride were extracted with chloroform-methanol (2:1, v/v) and the extracts were broken into two phases with 0.2 vol. of 0.1 M KCl.[1] One phase contained ^{3}H- and ^{14}C-labelled phospholipids and the other water-soluble ^{3}H- and ^{14}C-labelled metabolic intermediates.[2] Samples of these phases were used as test solutions.

Scintillation counting. The scintillator solution consisted of 7 g of 2,5-diphenyloxazole (PPO), 150 mg of 1,4-bis-(5-phenyloxazol-2-yl) benzene (POPOP) and 50 g of naphthalene dissolved in 1 litre of dioxan-toluene (4:1, v/v).

External standard channels radio method. A Beckman LS 100 scintillation spectrometer fitted with a cesium-137 automated external-standard source was aligned by using n-hexadecane standards in scintillator. The upper discriminator of window I was set so that the uppermost portion of the tritium spectrum was excluded (window I: tritium efficiency 39.50%; carbon-14 efficiency, 8.70%). The lower discriminator of window II was set just above the tritium energy spectrum and the upper discriminator just above the top of the carbon-14 energy spectrum (window II: tritium efficiency, 0.00%; carbon-14 efficiency, 66.20%).

The external standard was set up according to manufacturer's instructions to record a value of about 12.50 for the tritium standard.

Preparation of quench curves. Separate tritium and carbon-14 standards were prepared in duplicate and their radioactivities were counted to a 2σ error better than 0.2%. The external-standard channels-ratio value for each standard was obtained before and after counting and the average was recorded. The vials were then opened and a single drop of chloroform (about 0.02 ml) was added to each, the vials were resealed and the chloroform was mixed in by gentle swirling. After the standards had been placed in the dark for

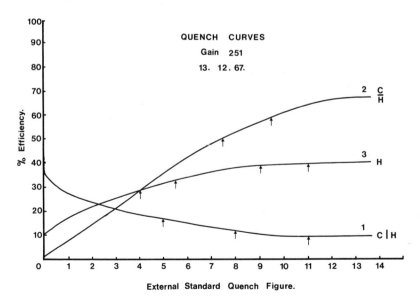

Fig. 1

5 min their radioactivities were recounted with the same precision. This procedure was repeated until the recorded external-standard channels-ratio had fallen to 0.00 (about 180 drops). The tritium and carbon-14 counting efficiencies in window I and the carbon-14 counting efficiencies in window II were calculated after each drop of chloroform had been added, and these were plotted against the respective average external-standard channels-ratio value (Fig. 1).

Hand calculations. The calculations necessary to separate the counts attributable to each radioactive isotope in binary-labelled samples and to determine d.p.m. per unit solvent volume are set out below. Counting efficiencies were read from graphs (Fig. 1) by applying the external-standard quench-ratio value printed out by the spectrometer. Subtraction of background counts:

$$(I)_p - k_1 = (I)_q \qquad (1)$$

$$(II)_p - k_2 = (II)_q \qquad (2)$$

Determination of carbon-14 counts in window I:

$$\frac{(II)_q \times y_1}{y_2} = R \qquad (3)$$

Determination of tritium d.p.m.:

$$\frac{[(I)_q - R]}{y_3} \times k_3 = D_t \qquad (4)$$

Determination of carbon-14 d.p.m.:

$$\frac{R + (II)_q}{y_1 + y_2} \times k_3 = D_c \qquad (5)$$

Tritium d.p.m. calculated per ml of solvent:

$$\frac{D_t \times k_4 \times v_2}{v_1} = \text{Tritium d.p.m./ml of solvent} \qquad (6)$$

Carbon-14 d.p.m. calculated per ml of solvent:

$$\frac{D_c \times k_4 \times v_2}{v_1} = \text{Carbon-14 d.p.m./ml of solvent} \qquad (7)$$

where $(I)_p$ is c.p.m. in window I (printed out by spectrometer), $(II)_p$ is c.p.m. in window II (printed out by spectrometer), $(I)_q$ is c.p.m. in window I with background counts subtracted, $(II)_q$ is c.p.m. in window II with background counts subtracted, k_1 is background counts in window I, k_2 is background counts in window II, k_3 is 100, k_4 is 1000 (factor converting d.p.m./mg into d.p.m./g), y_1 is carbon-14 counting efficiency in window I, y_2 is carbon-14 counting efficiency in window II, y_3 is tritium counting efficiency in window I, v_1 is sample weight (mg), v_2 is sample density, (g/ml) D_t is tritium d.p.m. (uncorrected for sample weight etc.), D_c is carbon-14 d.p.m. (uncorrected for sample weight etc.) and R is carbon-14 c.p.m. in window I.

Programmed calculations. An IME 86S desk-top calculator was used in conjunction with a programming unit (Muldivo Digicord DG 308) capable of storing 512 programme steps and an external data store (Muldivo MS 30) capable of storing 60 ten-digit values each with correct decimal point and sign. Values could also be summed into these stores. The results were printed out on an Imperial Data-log typewriter via a recoding output unit (Muldivo OP 207). Programming instructions can be entered manually through the calculator keyboard or inserted on eight-holed edge-punched cards. No special computer language is required.

Mathematical expression of quench curves

Any discrepancy between hand-calculated and computer-calculated results originates from errors made in determining counting efficiencies from mathematical expressions. Sufficient programme steps are available to fit to cubic or quartic polynomials, but it was decided that the curves shown in Fig. 1 could be adequately expressed by quadratic equations ($y = ax^2 + bx + c$, where y is the counting efficiency, x the external-standard quench-ratio and a, b, and c coefficients peculiar to each quench curve).

An entirely separate programme is used to determine the values of the coefficients a, b and c, which are subsequently entered into the data store for use during the calculation of results (see below). Programmes have been devised to calculate these coefficients by two different methods. The first is a comparatively simple programme, which derives them from the co-ordinates of three representative points on each curve (three-point method). The second takes a larger and variable number of points and obtains a line of best fit by the method of least squares (least-squares method). These methods are described here and the results compared.

Three-point method. The co-ordinates of three representative points on a curve are entered at the beginning of the programme (x_1, y_1; x_2, y_2; x_3, y_3). The programme then solves the following equations in sequence.

$$\frac{(y_3 - y_2)(x_2 - x_1) - (y_2 - y_1)(x_3 - x_2)}{(x_3 - x_2)(x_2 - x_1)(x_3 - x_1)} = a \qquad (8)$$

$$\frac{(y_3 - y_2)(x_2^2 - x_1^2) - (y_2 - y_1)(x_3^2 - x_2^2)}{(x_3 - x_2)(x_2 - x_1)(x_3 - x_1)} = b \qquad (9)$$

$$y_1 - ax_1^2 - bx_1 = c \qquad (10)$$

These are derived from the equations:

$$ax_1^2 + bx_1 + c = y_1 \qquad (11)$$

$$ax_2^2 + bx_2 + c = y_2 \qquad (12)$$

$$ax_3^2 + bx_3 + c = y_3 \qquad (13)$$

Least-squares method. The co-ordinates of a number of points on a curve are entered into a programme loop and values of Σx, Σx^2, Σx^3, Σx^4, Σy, Σxy, $\Sigma x^2 y$ and n are accumulated. When all co-ordinates have been entered, the programme passes from the loop to solve the following equations in sequence:

$$\frac{[(\Sigma x^3)^2 - \Sigma x^4 \Sigma x^2][\Sigma y \Sigma x^2 - \Sigma xy \Sigma x] + [\Sigma x^2 y \Sigma x^2 - \Sigma xy \Sigma x^3][(\Sigma x^2)^2 - \Sigma x^3 \Sigma x]}{[(\Sigma x^2)^2 - \Sigma x^3 \Sigma x]^2 - [(\Sigma x)^2 - n\Sigma x^2][(\Sigma x^3)^2 - \Sigma x^4 \Sigma x^2]} = c \qquad (14)$$

$$\frac{[\Sigma y \Sigma x^2 - \Sigma xy \Sigma x] + c[(\Sigma x)^2 - n\Sigma x^2]}{[(\Sigma x^2)^2 - \Sigma x^3 \Sigma x]} = a \qquad (15)$$

$$\frac{\Sigma xy - a\Sigma x^3 - c\Sigma x}{\Sigma x^2} = b \qquad (16)$$

These were derived from the following equations, which satisfy the principle of least squares (the 'normal' equations):

$$a\Sigma x^2 + b\Sigma x + nc = \Sigma y \qquad (17)$$

$$a\Sigma x^3 + b\Sigma x^2 + c\Sigma x = \Sigma xy \qquad (18)$$

$$a\Sigma x^4 + b\Sigma x^3 + c\Sigma x^2 = \Sigma x^2 y \qquad (19)$$

Calculation of results. The computer is programmed to follow the same sequence as that shown for the hand calculations. Sample number, c.p.m. in two channels (I_p and II_p) and the external-standard quench-ratio value are entered at the beginning of the programme and are first printed out as a permanent record. At the points in the programme where the three counting efficiencies are required (y_1, y_2 and y_3) the programme uses the appropriate coefficients and the external-standard quench-ratio value entered at the beginning of the programme to solve a quadratic equation. The nine coefficients (a, b and c for each curve), together with all constants, variables and intermediate results (shown above under 'hand calculations') are contained within the data store and are recalled into the calculations by the programming unit when required. Any value within the data store can be recalled between programme runs and altered if necessary. In this way, without reprogramming, the coefficients for the quench curves can be altered (thus changing the shape of the quench curves). Similarly, variables such as sample density or volume can be altered to accommodate different batches of samples. Finally, the programme prints out d.p.m. per unit solvent volume for each radioactive isotope, working registers are cleared and the machine returns to the start of the programme.

RESULTS
Accuracy of the quadratic equations in expressing the quench curves. The quench curves shown in Fig. 1 are not parabolic and thus cannot be fitted exactly by a quadratic

equation. Over much of their length, however, they are approximately so, the least parabolic portions being at each end of the curves. In our experience most samples fall within the external-standard quench-ratio range 3.00 to 11.00, and curves that accurately predict within this range, at the expense of the extremities, are generally to be preferred.

Three-point method. By selecting the three points away from the extremities it was possible to obtain a curve of acceptable fit. Table 1(a) compares efficiencies read from graphs by the eye with those calculated by the three-point method (arrows on the curves shown in Fig. 1 mark the points selected). It shows that computed efficiencies deviate from graph-read efficiencies by less than 0.5% efficiency over the range 3.50 to 11.50. Beyond these limits larger deviations occur. A better fit for the extremities of the curves could be obtained by selecting points spaced wider apart. Invariably, however, this resulted in less accurate reproduction of the central portion of the curve.

Least-squares method. Naturally this method produces a curve of best fit. If the co-ordinate values of points along the whole length of a curve are used then overall best fit is obtained. Table 1(b) shows that efficiencies derived by this method deviate from the graph-read values by less than 1.60% efficiency over the whole range shown. Deviations, however, over the middle of the quench-ratio range (3.50 to 11.50) are sometimes larger than by the three-point method (*cf.* Tables 1(a) and 1(b)). By sampling within the limits set for the three-point method (between the left-hand and right-hand arrows on each curve in Fig. 1) a closer fit over the more important middle portion of the quench curve is achieved. Table 1(c) compares the efficiencies obtained by sampling over the restricted range with those read from graphs. Now, a slightly better fit over the middle of the external-standard quench-ratio range occurs (*cf.* Tables 1(a) and 1(c)).

The improvement of fit by the least-squares method is relatively slight, and whichever procedure is used a decision must be made as to which portion of the curve must be most accurately predicted.

Computing system used to analyse radioactive hepatic phospholipids. Weighed samples (100 to 700 mg) of upper or lower phase of extracts of rat liver containing both tritium and carbon-14 isotopes were counted for radioactivity. Table 2 shows a selection of results obtained after computer simulation of the curves compared with those obtained by reading counting efficiencies by the eye from graphs of the type shown in Fig. 1. To minimise graph reading errors by the eye, a maximum and a minimum counting efficiency were taken for each external-standard quench ratio-value and a mean value was obtained by averaging the d.p.m./ml of solvent obtained for each efficiency. (Only the minimum efficiency is quoted in Table 2.)

Although percentage deviations are quoted in Table 2, this does not necessarily imply that the computed results are less accurate, especially when the deviations are small. The accuracy of the hand calculations is limited by the precision with which the curves are read by eye (say ± 0.1% efficiency), whereas the simulations are worked to several places of decimals. The final deviations in samples containing two radioactive isotopes are the result of the interaction of errors on each of the three curves used to separate tritium counts from carbon-14 counts in window I. In some instances errors for each curve have cancelled each other out, whereas in others they have become accumulated. In samples containing only one radioactive isotope the deviation is a direct function of the difference between a real and a simulated curve. The deviations quoted here for binary-labelled

Table 1. Counting efficiencies read from graphs by the eye compared with those computed from quadratic equations.[d]

External standard quench-ratio value	Carbon-14 counting efficiency in window I			Carbon-14 counting efficiency in window II			Tritium counting efficiency in window I		
	Graph value[a]	Computed value	Deviation	Graph value[a]	Computed value	Deviation	Graph value[a]	Computed value	Deviation
1(a) Three-point method									
0.50	30.70	26.54	4.16−	4.00	3.45	0.55−	14.00	15.23	1.23
1.50	25.00	23.93	1.07−	10.50	6.26	4.24−	19.10	19.63	0.53
2.50	22.00	21.52	0.48−	17.00	15.25	1.75−	23.10	23.41	0.31
3.50	19.40	19.29	0.11−	24.00	23.53	0.47−	26.40	26.79	0.39
4.50	17.20	17.25	0.05	30.80	31.09	0.29	29.40	29.75	0.35
5.50	15.40	15.40	0.00	37.50	37.94	0.44	32.30	32.30	0.00
6.50	13.90	13.74	0.16−	43.70	44.08	0.38	34.70	34.44	0.26−
7.50	12.30	12.27	0.03−	49.50	49.50	0.00	36.50	36.17	0.33−
8.50	10.80	10.98	0.18	54.20	54.21	0.01	37.70	37.49	0.21−
9.50	9.50	9.89	0.39	58.20	58.20	0.00	38.40	38.40	0.00
10.50	8.80	8.98	0.18	61.60	61.48	0.12−	38.90	38.90	0.00
11.50	8.60	8.27	0.33−	64.20	64.04	0.16−	39.20	38.99	0.21−
12.50	8.70	7.74	0.96−	66.20	65.89	0.31−	39.50	38.67	0.83−
1(b) Least squares method (best overall fit[b])									
0.50	30.70	29.36	1.34−	4.00	2.47	1.53−	14.00	14.41	0.41
1.50	25.00	25.90	0.90	10.50	10.50	0.00	19.10	18.84	0.26−
2.50	22.00	22.75	0.75	17.00	18.06	1.00	23.10	22.84	0.27−
3.50	19.40	19.92	0.52	24.00	25.13	1.13	26.40	26.40	0.00
4.50	17.20	17.40	0.20	30.80	31.72	0.92	29.40	29.53	0.13
5.50	15.40	15.21	0.19−	37.50	37.83	0.33	32.30	32.23	0.07−
6.50	13.90	13.33	0.57−	43.70	43.47	0.23−	34.70	34.50	0.20−
7.50	12.30	11.78	0.52−	49.50	48.62	0.88−	36.50	36.34	0.16−
8.50	10.80	10.54	0.26−	54.20	53.29	0.91−	37.70	37.74	0.04
9.50	9.50	9.62	0.12	58.20	57.49	0.71−	38.40	38.71	0.31
10.50	8.80	9.01	0.21	61.60	61.20	0.40−	38.90	39.25	0.35
11.50	8.60	8.73	0.13	64.20	64.93	0.73	39.20	39.36	0.16
12.50	8.70	8.76	0.06	66.20	67.18	0.98	39.50	39.03	0.47−
1(c) Least-squares method (restricted sample range[c])									
0.50	30.70	27.20	3.50−	4.00	2.71	6.71−	14.00	14.07	0.07
1.50	25.00	24.44	0.56−	10.50	6.65	3.85−	19.10	18.67	0.43−
2.50	22.00	21.88	0.12−	17.00	15.37	1.63−	23.10	22.80	0.30−
3.50	19.40	19.53	0.13	24.00	23.44	0.56−	26.40	26.46	0.06
4.50	17.20	17.38	0.18	30.80	30.86	0.06	29.40	29.65	0.25
5.50	15.40	15.44	0.04	37.50	37.63	0.13	32.30	32.37	0.07
6.50	13.90	13.71	0.19−	43.70	43.75	0.05	34.70	34.62	0.08−
7.50	12.30	12.18	0.12−	49.50	49.22	0.28−	36.50	36.40	0.10−
8.50	10.80	10.86	0.06	54.20	54.05	0.15−	37.70	37.71	0.01
9.50	9.50	9.74	0.24	58.20	58.22	0.02	38.40	38.55	0.15
10.50	8.80	8.83	0.03	61.60	61.75	0.15	38.90	38.72	0.02
11.50	8.60	8.12	0.48−	64.20	64.62	0.42	39.20	38.81	0.39−
12.50	8.70	7.62	1.08−	66.20	66.85	0.65	39.50	38.24	1.26−

a Values read by eye from the graphs shown in Fig. 1.
b Computed from co-ordinate points along the whole length of each curve.
c Computed from co-ordinate points within the limits set by the left-hand and right-hand arrows on each curve in Fig. 1.
d With permission from *Biochem. J.* **118**, 379 (1970).

Table 2. Comparison of results[b] obtained by 'hand calculation' and by desk-top computation[c].

Sample no.	External-standard quench-ratio value	Minimum counting efficiencies used in 'hand calculations'			'Hand-calculated' results		Deviations[a] by three-point method		Deviations[a] by least-mean-squares method	
		^{14}C counting efficiency in window I	^{14}C counting efficiency in window II	^{3}H counting efficiency in window I	10^{-3} x ^{14}C d.p.m./ml of solvent	10^{-3} x ^{3}H d.p.m./ml of solvent	^{14}C deviation	^{3}H deviation	^{14}C deviation	^{3}H deviation
1	12.08	8.70	65.90	39.40	14.33	362.74	1.30	2.11	0.44−	2.10
2	11.50	8.80	64.90	39.40	39.45	967.67	1.57	1.43	0.50	1.42
3	10.57	8.90	62.00	38.90	90.59	5646.90	0.65	0.10−	0.50	0.10−
4	9.49	9.60	58.10	38.50	43.55	1387.13	0.03−	0.07−	0.65	0.08−
5	8.48	10.80	54.20	37.70	45.24	1451.25	0.16	0.16	1.34	0.15
6	7.59	12.10	50.00	36.70	1712.23	2482.75	0.01	0.65	1.38	0.30
7	6.51	13.90	43.80	34.70	4.88	14.82	1.04−	0.90	0.16	0.64
8	5.77	15.10	39.70	33.00	4.69	14.04	0.28−	0.57	0.49	0.26
9	4.39	17.70	30.40	29.30	3.50	8.37	0.31−	0.54	1.63−	0.27
10	3.66	19.20	25.50	26.80	956.43	9877.37	1.29	0.13−	2.32−	0.08−
11	2.51	22.10	17.30	23.10	462.92	3644.65	9.35	0.76	2.71−	1.77
12	1.47	25.30	10.60	19.30	441.32	3614.15	59.87	4.42−	8.91	3.11
13	0.65	29.40	5.20	15.00	499.39	3881.33	576.34−	142.47	71.95	14.32

a Difference between d.p.m./ml obtained by the three-point method or the least-mean squares method and the 'hand calculations' expressed as a percentage of the hand-calculated value.
b The results are for samples of upper or lower phase from a chloroform-methanol extraction of rat liver containing ^{3}H- and ^{14}C-labelled phospholipids or their water-soluble metabolic intermediates. For the desk-top computation the least-mean-squares method giving overall best fit (as shown in Table 1(b)) was used.
c With permission from *Biochem. J.* 118, 379 (1970).

samples are from cases where both radioactive isotopes have been counted to a 2σ error better than 0.2% but have been considered as absolute values for the present purpose. If one isotope is counted to a smaller degree of precision than the other then obviously the errors incurred may be considerably increased.

Use of the system for channels-ratio quench correction. A series of samples was counted in a Nuclear Chicago Model 6801 spectrometer. The windows were set up using standards supplied by the manufacturer. Curves were constructed by plotting the ratio of counts in two of the channels against the counting efficiency. It was found that the system described above could simulate such curves with great precision for about 90% of their length. Only a slight modification in the programme used for calculating the results was necessary. First a channels ratio value was determined by dividing the sample activity in one channel by that in another. This value was then used in a similar way to the external standards channels-ratio value.

GENERAL CONCLUSIONS

Besides quench curves for chloroform-quenched (^{3}H)- and (^{14}C)-n-hexadecane samples in a dioxan-based scintillator (the most applicable to the samples counted here), a number of other quench curves have been constructed. These account for a range of radioactive isotopes and window settings, colour and chemical quenching agents and a toluene-based scintillator. As all curves can be adequately expressed by quadratic equations, reprogramming is not required. A record is kept of the computed coefficients

a, b and c for each set of curves, and these are fed into the data store before a batch of samples is processed.

The programmes described here have been drawn up to deal with radioactive isotopes of relatively long half-lives and no account has been taken of isotopic decay. Sufficient space remains, however, within the data store and programming unit for the decay calculations to be included if required. Also, data may be automatically reworked for samples channels-ratio standardisation, d.p.m. estimates made by this means may then be compared with estimates made by the external standard channels ratio method. A tape reader is now available for this computing system allowing it to accept directly the output from certain scintillation counters.

REFERENCES
1   J. Folch, M. Lees and G. H. Sloane-Stanley, *J. Biol. Chem.* **223**, 497 (1957).
2   G. H. Cope and M. A. Williams, *J. Microscopy* **90**, 31 (1969).

DISCUSSION

T. L. Woods: What is the average calculating time per sample (including data preparation time)?

M. A. Williams: For a reasonable number of samples (say 50), counted for 2 isotopes, it would average out at about 40 to 50 s each. This includes the two programs, the first to obtain the coefficients for the curves, the second to obtain d.p.m., subtract background, and relate d.p.m. to sample volume, density, weight, etc.

J. A. B. Gibson: The use of 180 points in establishing a quench curve is reasonable initially but there should be a simple relationship between curves. This should require a simple transformation of the variables (or axes) and hence only a limited number of points would be required for a repeat quenching curve.

M. A. Williams: Yes, it is sometimes quite possible to adjust the variables. However, at the start of a large batch of new material—every month or two—we prefer to check the curve. We feel that frequently people do not sufficiently define their quench curves, i.e. they do not use enough points.

D. S. Glass: What is the lowest efficiency you are prepared to work on for carbon-14? How important is it to calibrate at low efficiency levels?

M. A. Williams: We do not normally accept results from samples with total carbon-14 efficiencies below about 40%. Our first resort is to reprepare the sample. If this does not work we would consider doing some curve simulation, paying particular attention to the lower part (for double labelled samples). For single labelled materials one can always use an internal standard (one is not often driven to this!).

D. Moore: Have the calibration curves been produced for various quenching media as the shape of the curves will depend upon the material producing the quench?

M. A. Williams: Quench curves have been produced for a variety of chemical quenchers. The shapes of the curves vary, but in all cases simulation of the curves was feasible with this computing system.

C. P. Bond: With a similar instrument we find that for different counting systems, e.g. toluene as opposed to toluene/Triton X–100 (2:1), quench curves are entirely different—this may be due to the use of an emulsion system.

J. R. Clapham: What difficulties did you encounter due to the fact that data from the spectrometer is not on punch tape?

M. A. Williams: From our point of view the problems are small. We lose some time since we are tied to the computer during data processing. The time involved is about 30 to 40 min for 50 samples. To us this is not a serious matter.

B. Scales: Do your calibration curves vary in shape, depending on the scintillation mixture used?

M. A. Williams: Yes, they do. There are small changes in the general shape, together with changes in slope. Toluene based scintillators frequently give higher efficiencies than dioxan based ones. This, in itself, will alter the slope of the curves.

Chapter 9

A Comparison of Computer-Input Methods Used to Process Liquid Scintillation Counting Data

D. S. Glass and T. L. Woods

HOC Division, ICI Limited, Billingham, Teesside, England

INTRODUCTION

This chapter describes an investigation of different methods for processing liquid scintillation counting data available to our laboratory. The alternatives were assessed with respect to simplicity, speed, accuracy, presentation, storage and self-checking of results.

Our laboratory is a typical radiochemical laboratory attached to an industrial research department interested in petrochemicals of all types. A wide variety of radiochemical work is done using both carbon-14 and tritium as tracers for analytical and mechanistic studies. The laboratory operates a radiochemical counting service using two Packard liquid scintillation spectrometers (Models No. 3320 and 3314). Both carbon-14 and tritium are measured in consecutive batches of samples. The system to be described can accommodate dual labelled samples but we do very little of this type of work. We are fortunate that serious colour quenching is not often encountered. When it does arise, we try to avoid the problem by higher activities and dilution, where this is impractical, internal standard techniques are used.

The first part of the chapter describes the automatic calibration and quench correction procedures that we use, and the second part describes the data processing methods evolved to exploit them.

DESCRIPTION OF THE SYSTEM

The two liquid scintillation spectrometers are fitted with automatic external standards. A Westrex teletype is used for outputting data in visual and punched tape form from the Packard 3320 instrument, the older 3314 uses a Monroe digital printer. Counting conditions for each counter are shown in Table 1. These settings are arranged to give maximum sensitivity (E^*2/B) for carbon-14 and tritium in their respective channels, and to count the radium-226 standard with complete exclusion of carbon-14 in the third channel.

The scintillators used for routine analysis are described in Table 2. Samples are counted for preset time periods varying between 5 and 100 minutes as necessary to a maximum of 900000 counts. The external standard is counted for one minute after each sample.

Typical output from the teletype is shown in Table 3.

Table 1. Counting conditions

	Channel 1 (C1) Carbon-14		Channel 2 (C2) Tritium		Channel 3 (C3) Radium-226	
	Gain (%)	Window (v)	Gain (%)	Window (v)	Gain (%)	Window (v)
Packard 3314	8.0	50–1000	50	50–1000	3.0	350–1000
Packard 3320	5.6	50–1000	50	50–1000	2.0	350–1000

Table 2.

Aqueous scintillator		Organic scintillator	
10.0 gm	POP	4.0 gm	POP
0.25 gm	POPOP	0.1 gm	dimethyl-POPOP
100 gm	Naphthalene	1000 ml	toluene
700 ml	Dioxane		
300 ml	Ethanol		

Table 3. Typical output from the liquid scintillation spectrometer.

```
109  1445  562779  900000    119  866467  752612  124603
110  1413  900000  580655    121  849464  547644  240837
111  1353  900000  477457    355  832967  515749  268130
112  1448  641161  900000    114  874852  722677  139623
114  1597  900000  856507    132  891343  640330  209356
115  1283  900000  300712    133  744458  444420  322024
116  1491  727813  900000    129  880061  688389  160019
117  1290  900000  353376    131  771037  461613  303558
119  1458  512517  900000    114  840225  766546  108500
120  1457  900000  660666    117  869456  577572  232120
121  1279  900000  309804    130  742157  440587  313500
```

DATA PROCESSING FACILITIES

Over the past three years the data processing facilities available to us have improved considerably. Our earliest work in automatic quench correction was done with a Diehl Transmatic calculator.

An Olivetti desk-top computer then became available in the laboratory which revolutionised the approach to the work. A 'hands-on' facility in the form of a punched card reading IBM 1130 computer increased the computing power available as well as improving turn-round time for processing data by computer. Finally with the purchase of teletype output with the second liquid scintillation spectrometer we were able to use a 'hands-on' Elliot 903 computer reading paper tape and outputting data on tape or on the teletype.

CALIBRATION OF THE SYSTEM

The calibration system to be described makes it possible to measure and standardise each sample by two completely different methods. A description of the system and its applications have been reported.[1,2] Some of the advantages will be discussed later.

Calibration is performed using sets of quench sealed standards. Ampoules were prepared from low-potassium vials according to a suggestion of Dobbs.[3] The necks were drawn out and a glass tube attached through which the various materials could be added.

To each ampoule were added successively 10 ml of scintillator, 0.1 ml of spike solution (toluene-^{14}C for organic carbon, toluene-^{3}H for organic tritium and HTO for aqueous tritium) followed by varying quantities of a chemical quencher. Carbon tetrachloride was used for organic scintillator, water for aqueous scintillator. The ampoules were cooled in liquid nitrogen and then sealed under nitrogen.

Each set of standards was counted and the observed efficiencies and the corresponding channel ratios calculated. The ratios of interest are shown in Table 4.

Table 4. Channels ratios

	SCR	AES
Carbon-14	$\dfrac{\text{Net sample counts in C2}}{\text{Net sample counts in C1}}$	$\dfrac{\text{Net standard counts in C1}}{\text{Net standard counts in C3}}$
Tritium	$\dfrac{\text{Net sample counts in C1}}{\text{Net sample counts in C2}}$	$\dfrac{\text{Net standard counts in C3}}{\text{Net standard counts in C1}}$

For carbon-14 standards, a plot of observed efficiency against the sample channels ratio (SCR) defined as in Table 4, gives a smooth curve which can be fitted quite precisely by a quadratic model (Fig. 1). The inverted ratio gives a curve which can be fitted as precisely by a quintic polynomial (Fig. 2). Similar but certainly less smooth curves are obtained for the plot of the observed efficiencies against the external standard ratio (AES). The ratio (C3/C1) can be fitted reasonably well by a quintic polynomial, but the inverted ratio gives a curve which though straighter, is more kinked and cannot be fitted well by anything using a single model (Figs. 3, 4).

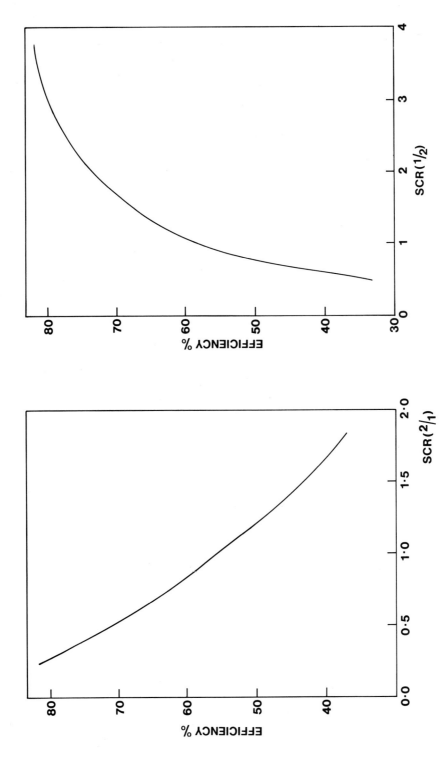

Fig. 1: Carbon-14 sample channels ratio (C2/C1) against efficiency.   *Fig. 2:* Carbon-14 sample channels ratio (C1/C2) against efficiency.

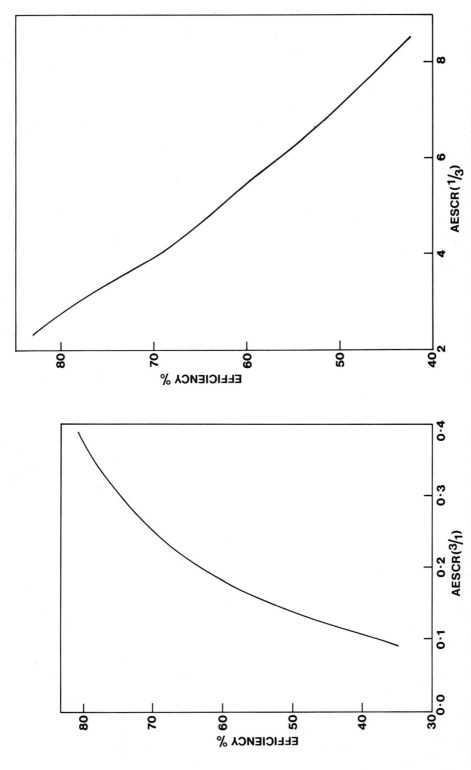

Fig. 3: Carbon-14 standard channels ratio (C3/C1) against efficiency. *Fig. 4:* Carbon-14 standard channels ratio (C1/C3) against efficiency.

The tritium SCR curves are smooth and can be fitted by quadratics or even linear models (Figs. 5, 6). The AES curve for organic tritium is curved at the top, but the aqueous tritium curve is practically linear until quite low efficiencies are reached (Figs. 7, 8).

CURVE MODELLING

Carroll and Hauser[17] have recently published an analysis of curve modelling and its relevance in liquid scintillation counting.

Linear and quadratic regressions to model the efficiency/ratio curves can be carried out rapidly by entering data manually on to the keyboard of an Olivetti desk-top computer. Polynomial regressions greater than quadratics are performed on the card-reading IBM 1130 using an IBM library program.

With the demand for greater accuracy in LSC results and the availability of 'hand-on' computing facilities, a program has been developed to fit the efficiency/ratio curves by multiple quadratic models using data direct from the teletype output of the counter. An improvement in fit is obtained by a curve-smoothing procedure. Each

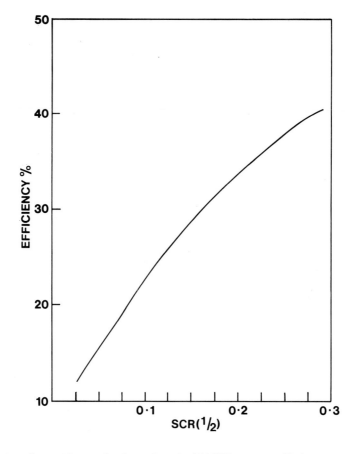

Fig. 5: Tritium (organic) sample channels ratio (C1/C2) against efficiency.

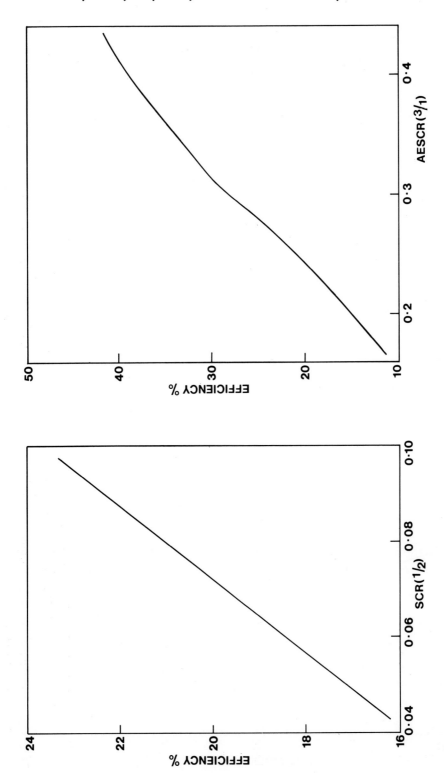

Fig. 7: Tritium (organic) standard channels ratio (C3/C1) against efficiency.

Fig. 6: Tritium (aqueous) sample channels ratio (C1/C2) against efficiency.

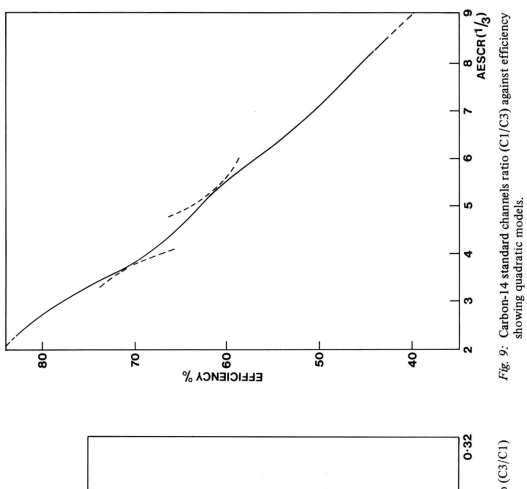

Fig. 9: Carbon-14 standard channels ratio (C1/C3) against efficiency showing quadratic models.

Fig. 8: Tritium (aqueous) standard channels ratio (C3/C1) against efficiency.

quenched standard produces an efficiency, a sample channels ratio and an external standard channels ratio. The quadratic model for the SCR/efficiency relationship gives predicted efficiencies for each observed sample channels ratio. The predicted efficiencies are then used together with the observed external standard channels ratios to obtain a series of quadratic models for the AES/efficiency curve. We find that three regressions sufficiently define the parts of the curve that we are interested in. Figure 9 shows how the three quadratic models (dotted lines) define the ^{14}C AES curve. This method defines the AES curves quite precisely since it removes scatter arising due to initial pipetting errors in making up the standards and some of the statistical errors in counting. The program that has been developed will do all this in a matter of seconds and print out the coefficients of the quadratics on isolated sections of punched tape to be used as described below.

The degree of fit used as a criterion for all these models is the residual standard deviation defined as shown in Table 5. Typical values for the residual standard deviation for the different polynomial models are shown in Table 6.

Table 5.

$$\text{Residual Standard Deviation} = \sqrt{\frac{\Sigma\ (\text{efficiency (observed)} - \text{efficiency (predicted)})^2}{\text{number of degrees of freedom}}}$$

Table 6. Residual standard deviations

Isotope	Efficiency Range (percent)	Deviations in percent efficiency using quadratic models	
		SCR	AES
Carbon-14 (organic)	30 to 85	0.27	0.90
	70 to 85		0.10
	59 to 71		0.21
	30 to 61		0.16
Tritium (organic)	15 to 55	0.69	0.95
	36 to 55	0.23	0.29
	15 to 36		0.11
Tritium (aqueous)	20 to 30	0.23	0.27

COMPUTER PROGRAMS

With the liquid scintillation counters set up to measure and calibrate each sample by two completely independent methods (AES and SCR), there was a pressing need for a fast and reliable data processing system to accommodate the volume of data which accumulates.

There have been a number of reports of computer programs written for processing liquid scintillation counting data. Table 7 gives a list of some of those described, as well as the quench correction procedures used and the methods of inputting data. These points are extremely relevant when considering available methods of data processing. Other points that we have found to be significant include the availability and accessibility of the computer, data turn-round time, the method of presentation of the results and the ease of storage of data.

The development of suitable programs was an object of this work, but the time spent on such development can be significant and must be proportionate to the amount of use made of the resulting system. The reported programs, however, were not relevant to our counting system and workable programs were quickly devised. Our present program has been updated many times from the original version. We have found that the speed, accuracy and versatility of the product easily justifies the work spent on it.

THE ELECTRIC CALCULATOR

This is of historic interest only.

Data is manually typed on to the keyboard and efficiences calculated using the simple storage facilities available. For polynomial relationships this can be an extremely long process. The mean time of calculation is approximately three minutes per sample for a quintic and due to the complexity of the operation the method is wide open to operator error. The calculator gives a printed output which must be copied and the method is of very little value except when the volume of data does not merit the outlay for more expensive equipment. Labour costs using such a system are high.

THE OLIVETTI DESK-TOP COMPUTER

There are many desk-top computers on the market. Scott[14] and Lumb[15] and Grower and Bransome[18] have published details of counters using an automatic key-punch system.

We use the Olivetti as an off-line, ready access computer for a wide variety of work which includes calculation of liquid scintillation samples.

Programs are written which can calculate channel ratios, activities and specific activities from raw liquid scintillation data in about thirty seconds. These programs are stored on magnetic program cards and programs are available for each isotope for each method of channel ratio/efficiency determination. Output is printed on a paper roll as a long list of figures (Table 8).

The computer is extremely simple to program, no special skills and little training being required. The programs we have written to solve quintic polynomials, the coefficients of which are written into the program, require about 100 instructions. We have also written programs using quadratic models, in which one SCR and two AES models produce two efficiencies and two specific activities (120 instructions). The results can be compared for consistency as described below.

For rapid calculation of results, for instance, in 'spot-checking' samples, the Olivetti is invaluable since the result can be obtained within seconds of the sample being measured. For larger numbers of samples, for instance, the product of a night's or weekend's counting, the relative merits of the ready-access Olivetti and a remote but larger computer become interesting. A series of ten samples takes about five minutes by Olivetti and is proportionately greater for larger numbers of samples. However, when calculating a series of samples at speed a number of input errors are liable to be made.

Table 7

Date	Computer	Language	Quench Correction	Input Method	Applications	Reference
1963	Royal McBee LGP-30	Act III	Internal Standard	Punched Tape (Direct)	Single Label Beta-emitting Isotopes	F. A. Blanchard[4]
	Burroughs-220	Algol	Internal Standard	Punched Tape (Direct)		
1965	IBM 7044	Fortran IV	Internal Standard	Punched Cards (Direct)	Single Label Beta-emitting Isotopes	J. L. Spratt[5]
1965	IBM 1620	Fortran	Discriminator ratio and external standard counts	Punched Cards (Direct)	Single and Double Label	L. R. Axelrod, et al.[6]
1966	IBM 7094	Fortran IV	Discriminator ratio	Punched Cards	Double Label (^{14}C and ^{3}H)	R. Ninomiya[7]
1966	IBM 7094	Fortran II/IV	Discriminator ratio	Punched Cards transferred from Punched Tape	Double Label	E. D. Plotka, et al.[8]
1966	IBM 1620	Fortran II	Discriminator ratio and internal standard	Punched Cards (Direct)	Double Label	C. Matthijssen[9]
1967	IBM 7044	Fortran IV	AES Counts	Punched Cards (Direct)	Single Label	J. L. Spratt, et al.[10]
1967	IBM 7094	Fortran	Sample channels ratio	Punched Cards (Manual)	^{89}Sr and ^{45}Ca	C. R. Creger, et al.[11]
1968	IBM 360/65	Fortran IV	AES ratio	Punched Cards transferred from Punched Tape	Double Label	J. T. O'Toole, et al.[12]
1968	Honeywell H-800	Fortran	Discriminator ratio	Punched Tape (Direct)	Single and Double Label	M. I. Krichevsky, et al.[13]
1968	Mathatron 850	Machine Code	Internal Standard	Punched Tape (Direct)	Single and Double Label	B. F. Scott[14]

Table 8. Olivetti output.

		V
Background Time (Min)	5 0	S
Background Counts (C1)	1 3 0 3	S
Sample Time (Min)	1 7 · 4 6	S
Sample Counts (C1)	8 3 2 7 3 5	S
AES Counts (C1)	9 0 9 8 6 6	S
AES Counts (C3)	1 3 6 1 2 2	S
AES Ratio (C3/C1)	0 · 1 5 7 8 8	A ◊
Efficiency (%)	5 4 · 9 2 1 7 8	A ◊
Activity (dpm)	8 6 7 9 3 · 4 1 5 4 5	A ◊
Sample weight (gm)	0 · 5	S
Specific Activity (dpm/gm)	1 7 3 5 8 6 · 8 3 0 9 0	A ◊
Sample Time (Min)	1 2 · 6 0	S
Sample Counts (C1)	9 0 0 0 0 0	S
AES Counts (C1)	7 9 1 4 9 4	S
AES Counts (C3)	3 2 3 9 1 1	S
AES Ratio (C3/C1)	0 · 4 4 9 8 3	A ◊
Efficiency (%)	0 2 · 0 3 2 2 2	A ◊
Activity (dpm)	8 7 0 4 2 · 2 6 5 7 2	A ◊
Sample weight (gm)	0 · 5	S
Specific Activity (dpm/gm)	1 7 4 0 8 4 · 5 3 1 4 4	A ◊

The limited storage capacity of the computer is a disadvantage, however. For instance, it is not possible to calculate simultaneously standard deviations and efficiencies by both methods using a single program on one side of the card. Lumb[15] has reported that it is possible to use a single card to do a double isotope analysis using quadratic polynomials and our calculation of two efficiencies using three quadratics is similar. Such programming, however, is only possible with care and maximum economy of space. Another drawback, particularly important in our 'service' work, is the manner of outputting the results. A long string of figures can be annotated and retained by the operator for reference, but for presentation in report form they must be transcribed into a suitable table. A larger computer will do this for you as described later.

The computer is a valuable facility in a laboratory in which a large amount of diverse calculation is done. Apart from calculating sample activities, we use the computer for decay corrections, mean and standard deviation measurements, and linear and quadratic regression calculations for curve fitting, and a wide variety of elementary operations. For these reasons we have not used the Olivetti either on-line with keyboard punch tied into the counter, or off-line with paper-tape input. Neither method would be compatible with the varied use we make of the computer. An off-line system with paper-tape input would have only slight advantage in speed over a more powerful paper-tape reading computer (if available) as described below and would not be able to cope with ancilliary data such as sample weights or individual background measurements without a lot of tape-editing.

IBM 1130 (PUNCHED CARD INPUT)

The two larger computers used in this comparison differ as far as we are concerned only in the manner of inputting and outputting data. Both the Elliott 903 and the IBM 1130 can use Fortran (or indeed the other computer languages) and both are 16k word machines. Both are available to us as 'hands-on' facilities.

The IBM 1130 available to us uses punched cards exclusively for input data, and a typewriter or punched cards for output data. The core size is quite adequate for all conceivable programs involving liquid scintillation counting data. Typical programs which we have developed require 1500 words with further array storage of about 1500 words (without the use of the COMMON area). Such programs can read in all data from cards, process it for efficiencies by both SCR and AES methods, calculate specific activities and standard deviations and output the results in a table which is self-explanatory and acceptable in most research reports and papers.

The rate determining step in our system, however, is the transfer of counter data to punched cards. Although a punching service is provided, the delays involved in such a service can sometimes be crucial. For counters which only print out data there is little to choose between punched-tape and card input since data has to be transferred anyway. The format and amount of data produced have considerable advantages over the Olivetti, at least for large numbers of samples.

Liquid scintillation counters can be linked directly to a card-punching machine. The necessary solenoid link is relatively inexpensive but a card-puncher is required and the rental or purchase of that is not cheap.

Data storage can become a problem since cards are bulky and can get out of order.

THE ELLIOTT 903 (PUNCHED-TAPE INPUT)

Since the teletype became available to us we have developed programs to accept punched-tape data directly from the counter, search for relevant data and calculate efficiencies and specific activities as before. One night's counting can be, and is, regularly completely processed and output in punched tape form in 10 to 20 min. (Some urgency is present in any case since the computer is used for other purposes 30 min. after the start of each day). The punched tape output is then printed up on the counter teletype at a convenient time.

Data is entered in three parts. The first section is a general tape for each isotope giving coefficients, valid efficiency ranges for the different quadratics for the AES curve and a trigger for the isotope. This is simply the output from the calibration program referred to earlier.

The second section is individual to each set of data and includes the number of samples, background and sample weight data. Sample vial numbers are given if the background or weight data are different for each sample. If they are the same for all or a part of the data, the relevant data is preceded by a zero instead of the sample number and it therefore applies to all the sample numbers not mentioned but encountered on the data tape.

The third section in simply the output from the liquid scintillation counter. This can be the whole tape from the previous night's counting since the second part defines the samples of interest (Table 9).

Table 9. Elliott input data.

```
    +1
     4
     0.3    2.0
     2.0    3.6
     3.6    5.5
     5.5    8.0
    -0.27239E+01   -0.23606E+02   0.90178E+02
    -0.16876E+01    0.16649E+01   0.85850E+02
    -0.12771E+00   -0.58943E+01   0.90849E+02
     0.95002E-01   -0.87547E+01   0.99893E+02
    'ELLIOTT OUTPUT DATA@
    10  1
        97   2000     489     562    183
        98   2000     424     439    163
        99   2000     463     613    165
       100   2000     444     549    179
         0   2000     467     533    177
        91   0.09910
        92   0.09830
        93   0.10790
        94   0.10540
        95   0.10070
        96   0.10000
         0   3.0
        91   2000   54168   24252   617  743337  458420  298134
        92   2000   83799   61790   402  758979  469181  277107
        93   2000   28824   13443   395  742799  455268  294610
        94   2000   29234   13576   542   76921  464482  298592
        95   2000   30123   12900   533  730466  450689  302921
        96   2000   30178   12943   572  724792  450839  301439
        97   2000    1711    1990   193  982072  697768  224954
        98   2000    1719    1990   212  972562  690013  226946
        99   2000     871     966   201  956894  684783  216974
       100   2000    1212    1247   225  971216  689097  222668
    -1
```

The feature of the input data is that the actual amount of information entered by hand is reduced to an absolute minimum and therefore reduces the possibility of input errors. At the very minimum nine figures need be entered together with the calibration and count data.

The output is designed to read as a self-explanatory table with messages to indicate inconsistencies or errors in the data. Each sample number is followed by its weight and the efficiency calculated by both AES and SCR methods. The specific activity is calculated using both efficiencies for comparison. Two standard deviations are calculated. The first is based solely on the number of sample and background counts assuming the efficiency is invariant. The second is derived from the error in the SCR efficiency based again on the number of sample and background counts. A simple check by the computer establishes whether the two specific activities are consistent in the light of these standard deviations. Providing there is more than one percent difference between the specific activities a check message is printed out if they are not consistent (Table 10).

The use of the double ratio method for checking consistencies in data is an extension of the double ratio method reported by Bush[16] and has been reported by us previously.[2]

Table 10. Elliott output data.

```
EFFICIENCY CALIBRATION CURVES, Y = A+B*X+C*X2

              RANGE (PERCENT)      A              B              C

SCR (C2/C1)   82.851 - 32.070    0.90178E+02   -0.23606E+02   -0.27239E+01
AES (C1/C3)   82.429 - 69.972    0.85850E+02    0.16649E+01   -0.16876E+01
AES (C1/C3)   67.974 - 54.567    0.90849E+02   -0.58943E+01   -0.12771E+00
AES (C1/C3)   54.616 - 35.936    0.99893E+02   -0.87547E+01    0.95007E-01
```

SAMPLE NO	SAMPLE WEIGHT	ACTIVITY CPM/GM	EFFICIENCY PERCENT AES/SCR	SPECIFIC ACTIVITY DPM/GM	STANDARD DEVIATION AES/SCR	COMMENTS
91	0.09910	27094.35	79.57 79.22	34051.17 34201.33	117.93 39.37	
92	0.09830	42386.57	77.86 71.35	54437.53 59403.70	147.65 90.59	CHECK THIS RESULT
93	0.10790	13140.41	79.35 78.87	16559.64 16661.61	79.31 27.52	
94	0.10540	OUT OF RANGE, RATIO=0.2527 ACTIVITY= 13646.58 CPM/GM				
			78.92	17292.75	28.24	CHECK THIS RESULT
95	0.10070	14724.93	80.08 79.86	18387.23 18438.37	86.84 27.59	
96	0.10000	14855.50	80.13 79.84	18539.86 18605.94	87.53 27.85	
97	3.00000	20.37	62.68 58.87	32.49 34.59	0.78 1.07	
98	3.00000	21.58	63.25 58.00	34.13 37.21	0.77 1.11	
99	3.00000	6.80	62.37 67.72	10.90 10.04	0.61 0.52	
100	3.00000	12.80	62.71 66.47	20.41 19.26	0.68 0.61	

It is used for checking homogeneity, chemiluminescence, misprinting, instrument drift and malfunction, and other errors.

A computer system capable of checking its own results has tremendous advantage when large numbers of samples are processed.

A disadvantage of the data processing system as it is set up is the difficulty of amending data arising due to misprints. This is, however, a minor point. A comparison of processing times using the Olivetti, the Elliott and the IBM systems shows a clear advantage to the punched-tape system for large numbers of samples. The Olivetti on the other hand is more suitable for small numbers. This is illustrated in Fig. 10. The break-even point for the Olivetti and the Elliott depends on the presentation and the amount of information required. For less than ten samples there may be no advantage in assembling the data tape, although this can always be done before the samples are counted. The Olivetti, then, will give a rapid and sufficient result. If large numbers of samples are encountered, however, and considerations of presentation, consistency and storage enter in, the punched-tape system using the Elliott is unparalleled.

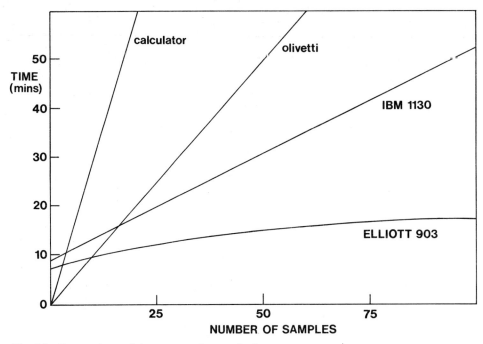

Fig. 10: Comparison of data processing methods.

REFERENCES

1. D. S. Glass, *Intern. J. Appl. Radiation Isotopes* **21**, 531 (1970).
2. D. S. Glass, Proceedings of the Second International Symposium on Organic Scintillators and Liquid Scintillation Counting, San Francisco, California, USA, 5th to 9th July 1970, Academic Press, New York, in press.
3. H. L. Dobbs, personal communication.
4. F. A. Blanchard, *Intern. J. Appl. Radiation Isotopes* **14**, 213 (1963).
5. J. L. Spratt, *Intern. J. Appl. Radiation Isotopes* **16**, 439 (1965).
6. L. R. Axelrod, C. Matthijssen, J. W. Goldzicher and J. E. Pullman, *Acta Endocrinol. Suppl.* 99 (1965).
7. R. Ninomiya, *Intern. J. Appl. Radiation Isotopes* **17**, 355 (1966).
8. E. D. Poltka, E. G. Stant, Jr., F. A. Waltz, V. A. Garwood and R. E. Erb, *Intern. J. Appl. Radiation Isotopes* **17**, 637 (1966).
9. C. Matthijssen, *Anal. Biochem.* **15**, 382 (1966).
10. J. L. Spratt and G. L. Lage, *Intern. J. Appl. Radiation Isotopes* **18**, 247 (1967).
11. C. R. Creger, M. N. A. Ansari, J. R. Couch and L. B. Colvin, *Atompraxis* **13** (1), 24 (1967).
12. J. J. O'Toole and J. O. Osborn, *Intern. J. Appl. Radiation Isotopes* **19**, 821 (1968).
13. M. I. Krichevsky, S. A. Zaveler and J. Bulkeley, *Anal Biochem.* **22**, 442 (1968).
14. B. F. Scott, *J. Radioanal. Chem.* **1** (1), 61 (1968).
15. B. R. Lumb, *World Med. Instrumentation* March 1969, p. 8.
16. E. T. Bush, *Intern. J. Appl. Radiation Isotopes* **19**, 447 (1968).
17. C. O. Carroll and T. J. Hauser, *Intern. J. Radiation Isotopes* **21**, 261 (1970).
18. M. F. Grower and E. D. Bransome, *Anal. Biochem.* **23**, 159 (1969).

DISCUSSION

D. Smith: With access to a large computer store, I wondered why you do not count your standards and calculate the curve constants for every counting run?

D. S. Glass: We check our standards every weekend. Running 20 and 50 samples a night leaves us no time to count standards for every run. In fact, the stability is such that recalibration is not required very often. In addition, the self-checking procedure immediately diagnoses instrument drift.

P. Hill: To all intents the Muldivo system can offer similar facilities to that of the Elliot 903. With the availability of formating, print-out on to a preprinted sheet alleviates the problem of transcribing results into a readable form. There is less storage and other such facilities yet check systems can be introduced in the Muldivo system to give the same facilities 'on check' as the 903 without, of course, alpha output.

D. S. Glass: Possibly—but not as quickly and, as you remark, without any alphanumeric display. This leads to less versatility and clarity in presenting results. In addition, I would have thought that the desk-top computing system would be unable to cope with a long list of sample weights or individual background measurements due to its limited storage capacity. If these ancillary data had to be entered, together with the respective pieces of count data, the preparation of the data tape would be awkward and time-consuming. However, the preference has to lie with a combination of the individual user's requirements and the demand for local computer facilities. A large computer with its versatility, speed and better storage facilities may be justified if fully utilised and accessible. A smaller but still expensive desk-top unit must have its place in establishments where computer facilities are less in demand, but in which significant islands of activity exist.

Chapter 10

A New Gelifying Agent in Liquid Scintillation Counting

A. Benakis

Laboratory of Drug Metabolism,* University of Geneva, Geneva, Switzerland

Several techniques for preparing samples for liquid scintillation counting have been described in the literature. Solid samples of materials (i.e. $Ba^{14}CO_3$, powdered biological material, absorbant of thin-layer chromatography plates, inorganic salts, etc.) that are not soluble in liquid scintillation mixtures may be counted by suspending the material as a powder in the liquid scintillant.

Originally the counting sample was simply shaken and then counted, allowance being made for the settling that occurred. This technique was not valuable because the efficiency and reproducibility was not satisfactory.

More satisfactory results are obtained, however, when a gelifying agent is incorporated in the liquid scintillant to prevent settling. The material is homogeneously suspended in the vial containing the liquid scintillator.

Aluminium stearate has been used at a concentration of 5% in toluene or in xylene,[1] containing the fluors (PPO, POPOP). The aluminium stearate, however, has the disadvantage of making a hard gel, creating a difficult homogeneous distribution of the solid radioactive sample.

Another material, named 'Thixcin' (ricinoleic acid), has been used as a gelifying agent.[2] This material enables the preparation of a certain quantity of gelifying gel in advance, at a concentration of 25 g of 'Thixcin' in powder for 1 litre of scintillating liquid. The samples of the liquid scintillation prepared with this product have a good optical transmission. Nevertheless, its use as a gelifying agent has some disadvantages, in particular because it loses its gelifying properties at room temperature, allowing the suspended product to deposit.

Another product has been described in the literature and used, namely colloidal silica.[3] It has been used under the name of 'Cab-O-Sil'† at a concentration of 4% by

---

*Pavillon des Isotopes, 20 Bd. d'Yvoy, Geneva, Switzerland.
†Trademark of Cabot Corporation, USA.

weight in the toluene containing fluors. It is widely used for measuring the radioactivity of solid material suspended in gels. For quantities of 100 to 200 mg in 15 ml of scintillating solvent, it does not show any effect on light transmission and can bear quantities up to 1 g of suspended material. Nevertheless, 'Cab-O-Sil' has several disadvantages, like the difficulty in handling when weighing, this material being very bulky and very sensitive to electrostatic charges. Furthermore, at room temperature it is inclined to let the suspended material settle. On the other hand, in certain applications the silica, because of its great reactivity, absorbs polar radioactive material, thus diminishing the efficiency of the measure of the radioactivity by liquid scintillation.

Another gel scintillator has been described.[4] It is formed by adding toluene diisocyanate to a toluene cocktail containing branched aliphatic amines. The toluene diisocyanate used is Hylene TM–65 (Dupont) (mixture approximately 65% toluene 2,4-diisocyanate and 35% toluene 2,6-diisocyanate). Armeen L–11 (Armour Ind. Chem. Co.) was used as the source of branched aliphatic primary amine.

The gel scintillator described in this chapter consists of polyolefine resins which are the result of the polymerisation of ethylene, propylene, butylene and similar products as well as of the co-polymerisation of two or more of these olefines with low molecular weight.

This product is known under the name of 'Poly-Gel-B'.[5] It is in the form of fine opalescent white pellets. Poly-Gel-B pellets are soluble in solvents such as toluene, xylene, etc. when the solvent is heated at a temperature of 60 to 70°C. After cooling at room temperature a homogeneous gel is formed. This product already contains primary and secondary fluors like PPO/POPOP.

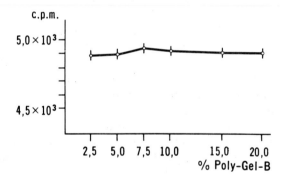

Fig. 1: Concentration effect of Poly-Gel-B in counting efficiency of 50 mg $Ba^{14}CO_3$.

It is currently used at a rate of 10% in the solvent. Figure 1 shows that the concentration has no effect on the counting efficiency of $Ba^{14}CO_3$. The concentration depends on the material to be suspended. Our experiment, however, showed that this 10% quantity is sufficient for $Ba^{14}CO_3$ and for powder of biological material. As shown in this figure, an increase of Poly-Gel-B does not decrease the counting efficiency. If it is necessary to suspend high density material the concentration can be increased.

For one litre of toluene, 100 g of pellets of Poly-Gel-B should be used. The solvent is heated in a water bath to a temperature of 60 to 70°C for a few minutes until the gelifying agent is completely dissolved. The quantity of solvent containing the gelifying agent is placed hot in the scintillation vial already containing the material to be suspended. One can also add pellets of Poly-Gel-B in amounts of 10% or more directly in the scintillation vial already containing the toluene. The vial, even if capped, is then heated for about five minutes until the gelifying agent completely dissolves. The gel is formed when the solution is brought back to room temperature. In case a quick formation of gel is required, the scintillation vial can be cooled in ice. The gel can be put back into solution by simply heating the vials.

Several radioactive isotopes in suspension can be counted under the form of a gel, i.e. carbon-14 in the form of $Ba^{14}CO_3$ obtained after combustion of biological material, sodium carbonate after precipitation, carbon-14 contained in the powder of biological material, plant material, absorbant of thin-layer chromatography plates like cellulose powder, alumina powder, etc., tritiated compounds as contained in organic material or on thin-layer chromatography absorbant.

Other isotopes can be counted in suspension, i.e. iron-55 or iron-59 in the form of insoluble white ferri-phosphate complex or benzene phosphinate complex, calcium-45, strontium-90 in the form of sulphate, caesium-137 as perchlorate, iodine-131 in the form of silver iodide.

We have compared the counting efficiency of $Ba^{14}CO_3$ using Cab-O-Sil and Poly-Gel-B. Figure 2 shows that Poly-Gel-B is as efficient as the Cab-O-Sil. It even slightly increases the counting efficiency.

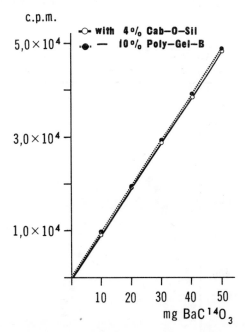

Fig. 2: Comparison of results obtained by suspension scintillation counting of $Ba^{14}CO_3$ with Cab-O-Sil and Poly-Gel-B.

Figure 3 shows a comparison of scintillation counting of rat liver powder in toluene and in suspension in Poly-Gel-B. For quantities over 20 mg, one can see a notable increase of the counting ratio.

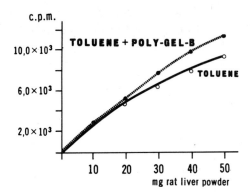

Fig. 3: Comparison of results obtained by scintillation counting of biological material (rat liver powder) in toluene and in suspension with Poly-Gel-B.

The samples thus prepared have shown a perfect suspension without any settling, even after several months at room temperature. This stability at room temperature is important as several liquid scintillation counting spectrometers work at room temperature.

Poly-Gel-B does not show any effect of light transmission and has no quenching effect. The addition of Cab-O-Sil and Poly-Gel-B does not modify the spectra of toluene carbon-14 measured in a Beckman Spectrometer LS–200B in the window-range of 0 to 400 (Fig. 4).

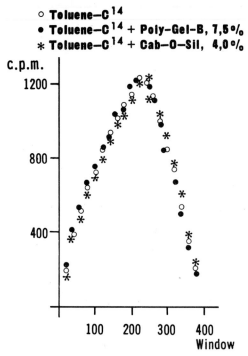

Fig. 4: Effect of Cab-O-Sil and Poly-Gel-B in carbon-14 spectra of homogeneous internal standard of toluene-^{14}C.

Fluors other than PPO/POPOP can be used, i.e. butyl-PBD* associated with the POPOP. As indicated in Fig. 5, there is a slight increase of the counting efficiency of the $Ba^{14}CO_3$ when fluor butyl-PBD/POPOP is used. In contrast, the butyl-PBD alone does not give such good results.

$BaC^{14}O_3$		PPO/POPOP		Butyl-PBD		Butyl-PBD/POPOP	
mg	dpm	cpm	Eff.%	cpm	Eff.%	cpm	Eff.%
10	10 500	9 394	89,4	8 859	84,3	9 591	91,3
20	21 000	18 847	89,7	18 839	89,7	18 924	90,1
30	31 500	28 077	89,1	28 275	89,7	27 680	87,8
40	42 000	37 561	89,4	37 430	89,1	37 022	88,1
50	52 500	47 323	90,1	47 982	91,3	46 835	89,2

Standard material: $BaC^{14}O_3$ ($1,05 \times 10^3$ dpm/mg)

Scintillation cocktail:
— toluene, PPO/POPOP (4/0,05 g/l), 10% Poly-Gel-B
— toluene, butyl-PBD/POPOP: (7/005 g/l), 10% Poly-Gel-B

Fig. 5: Counting efficiency of different amounts of $Ba^{14}CO_3$ in a Poly-Gel-B gel scintillator with various fluors.

The f value described as the ratio of suspension-counting efficiency to homogeneous internal-standard counting efficiency is determined by using various quantities of $Ba^{14}CO_3$. Figure 6 compares the influence of Cab-O-Sil and Poly-Gel-B on this f value. In fact, the f value of the $Ba^{14}CO_3$ is better with Poly-Gel-B than with Cab-O-Sil.

	f VALUES		
$BaC^{14}O_3$ mg	CAB-O-SIL PPO/POPOP	POLY-GEL-B PPO/POPOP	POLY-GEL-B Butyl-PBD/POPOP
100	0,95	0,99	0,98
200	0,94	0,97	0,98
300	0,93	0,96	0,94
400	0,91	0,95	0,94
500	0,90	0,95	0,91

f : ratio of suspension-counting efficiency to homogeneous internal-standard counting efficiency.

Standard material: $BaC^{14}O_3$, (105 dpm/mg)
Scintillation cocktail:
— toluene, PPO/POPOP (4/0,05 g/l), 4% Cab-O-Sil
— toluene, PPO/POPOP (4/0,05 g/l), 7,5% Poly-Gel-B
— toluene, butyl-PBD/POPOP (7/0,05 g/l), 7,5% Poly-Gel-B

Fig. 6: Suspension f values for different amounts of $Ba^{14}CO_3$ in Cab-O-Sil and Poly-Gel-B.

---

*2(4'-t-butylphenyl)-5-(4''-biphenyl)-1, 3, 4-oxadiazole

Biological material can also be counted in suspension and absolute values can be determined if calibration curves have been prepared in advance. Figure 7 shows that the counting efficiency depends on the volume of scintillation liquid and on the amount of biological powder. When rat liver powder is counted, one can obtain a relative efficiency of about 44% in comparison after determination with oxygen flask combustion. This figure shows also that the quenching calibration factor S is better in the case of measurements of 50 mg in 20 ml of liquid scintillation solution.

Rat liver powder mg	liquid scintillation solution: 15 ml		liquid scintillation solution: 20 ml	
	S	Relat. Eff. %	S	Relat. Eff. %
50	0,542	43,5	0,711	44,5
50	0,599	43,3	0,732	43,9
100	0,388	37,7	0,479	37,1
100	0,336	35,6	0,497	37,3

Scintillation cocktail:
– toluene, PPO/POPOP(4/0,05 g/l), 10% Poly-Gel-B
– toluene, butyl-PBD/POPOP(7/0,05 g/l), 10% Poly-Gel-B
S: quench calibration factor

Powder of rat liver containing C^{14}:
– Radioactivity of 100 mg: 86 900 cpm
    measured after oxygen flask combustion.

Fig. 7: Counting efficiency of biological material in suspension counting with Poly-Gel-B in various experimental conditions.

The gelifying agent Poly-Gel-B must not be used with cocktails of the Bray type containing dioxane. It can, however, be used with toluene containing 10% ethanol. In the case of measurements of radioactivity of biological material like urine, plasma, etc., another type of gelifying agent can be used, Poly-Gel-B13, which allows the formation of an emulsion.

In conclusion, the use of Poly-Gel-B in liquid scintillation counting as a gelifying agent, presents the following advantages:
(a) The gels obtained are suspended without any settling even at room temperature.
(b) Numerous materials can be gelified provided calibration curves are prepared and absolute measurements can be made.
(c) It allows counting of material of various densities.
(d) These gels prevent some labelled compounds such as lipids from being absorbed on the glass surface of the scintillation vial, thus increasing the counting efficiency.
(e) This gelifying agent does not cause a quenching effect and decline of counting efficiency of ^{14}C ^{3}H labelled substances.
(f) This material is easily handled. It is of low cost and its properties should make the switch from other currently used gel scintillators to Poly-Gel-B easy and advantageous.

ACKNOWLEDGEMENT
The author wishes to thank Miss J. Corthay and Mr. C. Gachet for their technical assistance.

REFERENCES

1. B. L. Funt, *Nucleonics* **14** (8), 83 (1956).
2. C. G. White and S. Helf, *Nucleonics* **14** (10), 46 (1956).
3. D. G. Ott, C. R. Richmond, T. T. Trujillo and H. Foreman, *Nucleonics* **17** (9), 106 (1959).
4. J. N. Bollinger, W. A. Mallow, J. W. Register Jr. and D. E. Johnson, *Anal. Chem.* **39**, 1508 (1967).
5. A. Benakis, French Patent No. 1,590,762. Other patents, including US, G.B., Japan, Germany and Canada, pending.

DISCUSSION

D. A. Kalbhen: (i) In my own experiments with your Gel I never got counting efficiency above 8% with coloured biological materials, such as red blood cells or liver. How can you get 45% efficiency with liver powder? (ii) What is the advantage of Poly-Gel-B for samples after oxygen combustion?

A. Benakis: (i) I expect that you will not obtain better results if you suspend your red blood cells in Cab-O-Sil! Your results show well that the labelled compound is absorbed by the red blood cells and, under such conditions, a combustion would be preferable.
(ii) It still happens in literature references, that people make combustions and after it obtain barium carbonate ^{14}C. In such cases Poly-Gel-B can be really useful in the barium carbonate counting techniques.

A. R. Ware: Regarding some results given by the author using $Ba^{14}CO_3$ standard in which he claims 90% efficiency and extremely good reproducibility, I would like to hear the author's comments on the high efficiency and, more particularly, his views on the effects of self-absorption, etc., caused by variation in the particle size of the suspended material.

A. Benakis: The barium carbonate we have been using was supplied by New England Nuclear Corporation, Boston. It was prepared for instrument calibration. Its specific activity was 1.05×10^3 d.p.m./mg. The high efficiency may be attributed to the physical structure of barium carbonate. In fact, it is known that the barium carbonate counting efficiency in liquid scintillation counting depends on how the product was precipitated. The way barium carbonate is precipitated influences its crystallisation. On the other hand, the barium carbonate difference in counting efficiency was brought to our attention by persons who had been using the suspension technique.
Our results are based on the radioactivity values given by New England Nuclear Corporation. We believe that these values are exact, but we did not have the possibility of controlling them by another technique. I am of the opinion that the barium carbonate technique may be used, but the precipitation conditions should be rigorously established and it is preferable to check the radioactivity of this material with another technique, like by formation of CO_2 and counting in a proportional counter.

B. W. Fox: Comment: I am surprised that the levels of efficiency quoted for this system are so high. I would expect an efficiency of about 40% due to self-absorption factors. Question: Do you have any experience of measuring tissue homogenates in this system?

A. Benakis: I answered Dr. Ware as far as the high efficiency level is concerned. Of course, the efficiency depends on the quantity of barium carbonate suspended. This is controlled by self-absorption. Self-absorption depends on the quantity and also on the crystalline structure of the barium carbonate.

As to your question on tissue homogenates, we have experience with organic powders, as indicated in the report, but not with tissue homogenates, because Poly-Gel-B is not compatible with aqueous solutions.

Another gelifying agent, Poly-Gel-B13, is used to form emulsions. We are presently perfecting another product: Poly-Gel-H, specially for use with cocktails of the Bray type, containing dioxane. Nevertheless, considering the pigmentation of the tissue homogenates, it is not recommended to measure them without first discolouring them.

Chapter 11

Measurement of Radiation Effects on Thyroid Cell DNA Synthesis using Tritiated Thymidine

William R. Greig

*University Department of Medicine and Nuclear Medicine,
Royal Infirmary, Glasgow, C.4., Scotland*

INTRODUCTION

The rat thyroid is a suitable model with which to study radiation effects on thyroid cell proliferation *in vivo*. The model has the advantage of a relatively simple cell population; about 70% of the cells are follicular and about 30% are stromal.[1] There is no migration of cells to or from the gland either normally or when growth is artificially promoted by a goitrogenic challenge. The tissue is thus normally closed and non-dividing, or when goitrogen stimulated which induces cell hypertrophy and hyperplasia it is closed and dividing.[2]

Ionising irradiations impair the capacity of the rat thyroid to undergo its normal 2 to 3 fold increase in weight in response to a continuous goitrogenic growth stimulus *in vivo*. The degree of impairment of weight response has been employed as an approximate index of irradiation effects on the reproductive potential of thyroid cells.[3,5] In these latter studies an assessment was made of changes during growth with and without irradiation due to cell size and number using histological measurements. In the current investigation these studies are complemented by biochemical measurements of nucleic acid synthesis. In addition irradiation on the follicular and stromal cells have been distinguished in the present study.

A detailed study of the effects of different doses of X-irradiation on the thyroid cell population of the normal rat and on the cell population during goitrogenic growth was carried out. The techniques employed included sequential measurement of total thyroid weight, cell density, cell composition, RNA and DNA synthesis using chemical and cell labelling methods. The data are critically discussed with emphasis on how to use the rat thyroid as a radiobiological model with which to study radiation effects on proliferation of differentiated thyroid cells *in vivo* and DNA synthesis too. X-irradiation was used as a precisely dosed and homogeneous radiation.

METHODS AND MATERIALS

X-irradiation. All the methods are standard; X-irradiation of the rat thyroid was through the ventral neck surface of animals as described by Crooks et al.[6] The conditions were 300 KeV, 20 mA, 23 cm FSD, and the dose rate was 190 rad per min. Control animals were anaesthetised but were not irradiated. All rats were adult male Sprague-Dawley common stock (Tuck and Sons, England).

The rats were sacrificed using coal gas or ether and each freshly dissected thyroid lobe was weighted to 0.1 mg. Thyroid weights were expressed as the mean weight of the two thyroid lobes for the groups specified below.

Cell density and composition. Haematoxylin and eosin sections ($5\,\mu$) of thyroid were prepared (in a research pathology laboratory) from one lobe and viewed through a squared eyepiece graticule using conventional light microscopy (Watson Microsystem 70). The magnification and viewing characteristics were kept constant and the total number of cells, follicular and stromal, in fifty fields were counted. Relative cell density was expressed as the average number of thyroid cells of all types per field. In the same fields the follicular cells were distinguished from the stromal cells and the percentage ratio of each was obtained.

Chemical RNA and DNA. The other thyroid lobe was stored at $-20°C$ and the total DNA and RNA estimated exactly as published by Begg, McGirr and Munro.[7] These methods are based on perchloric acid extraction and alkali digestion; DNA was measured by the Ceriotti colour reaction and RNA by optical density at $260\,\mu$. Mean total DNA per thyroid and mean total RNA per thyroid for the rats in each group were expressed in μg. The DNA control was calfthymus DNA (Sigma/type 1), and RNA control was purified yeast (Sigma/type X1).

DNA Synthesis. When DNA synthesis was to be studied in a pulse label of tritiated thymidine (Thymidine-6-^{3}H Radiochemical Centre, Amersham) was given intraperitoneally in a single dose of 0.5 μCi per gram body weight. The specific activity varied from 17.0 to 28.0 Ci/mM. The animals were killed 4 hours after injection and thyroids from each treatment group were pooled. Preliminary nucleic acid extraction was made as described by Begg, McGirr and Munro.[7] The precipitate containing the DNA was digested using 0.6 N Nuclear Chicago Solubiliser in toluene.[8,9] Counting was carried out in the Toluene-PPO-POPOP liquid scintillation system[10] and ultimately expressed as disintegrations per minute (d.p.m.) Tritiated thymidine incorporation into the nuclei of thyroid cell populations was the mean d.p.m. per whole thyroid for rats in each treatment group as specified below. Chemical DNA was not measured simultaneously with tritiated thymidine since it was found in preliminary experiments that the complete chemical procedure resulted in loss of label occurring at stage of alkali digestion.

DNA labelling index. Nuclear emulsion autoradiographs were also prepared using Kodak N.T.B. 2 nuclear emulsion exactly as described by Kopriwa and Leblond.[11] They were exposed at $4°C$ for three to four weeks, stained with neutral red, mounted and viewed using the microscopic characteristics described above for the determination of cell density. Fifty fields were surveyed from each treatment group and the number of cells labelled were counted. Labelled cells were those with a minimum of 20 grains. Mirror sections not used for autoradiography were stained with haematoxylin and eosin and the total number of cells in the 50 fields were determined. The labelling index was expressed as the number of labelled cells per 5000 cells. In all cases the labelled cells were randomly distributed throughout the tissue sections.

EXPERIMENTS, RESULTS AND INTERPRETATIONS
Experiment 1: Effects of X-rays on normal thyroid.

Thyroid weight, cell density, chemical DNA and RNA, tritiated thymidine incorporation and labelling index were determined at regular intervals before and after a single dose of 500 rad, a large dose in radiobiological terms. Groups of six rats were sacrificed at each of the nine times indicated.

The data (Fig. 1) demonstrate that in normal adult rats there is virtually no cell

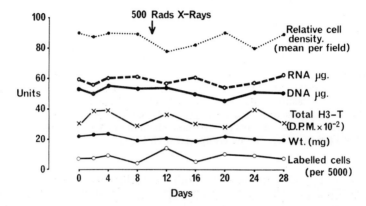

Fig. 1: Cell proliferation and DNA synthesis in normal rat thyroid. Sequential measurements on normal rat thyroid before and after single dose of X-rays (500 rad). Values are means from 6 animals per sacrifice.

proliferation and that a single X-ray dose of 500 rad induces no changes; the potentially most sensitive indices of normal cell proliferation, tritiated thymidine incorporation and labelling index, both remained very low throughout the period of observation.

It can be concluded that the adult rat thyroid cell population is stationary in phase G_0 or G_1[2] and that X-irradiation at a dose of 500 rad does not produce observable change. The next experiments describe the detailed changes in cell proliferation and DNA synthesis brought about by a goitrogenic challenge *without irradiation*; these provide control data for reference when the thyroids were X-irradiated then cell proliferation was promoted by the goitrogenic challenge.

Experiment 2: Effects of a goitrogenic challenge on unirradiated thyroid.

Thyroid growth was artificially stimulated by providing 0.1% aqueous methylthiouracil (4-methyl-2-thiouracil, BDH Labs.) as drinking water and a diet of low iodine content (Nutritional Biochemical). Groups of five rats were sacrificed at each of the seven times indicated in Fig. 2, during which intervals the drug interrupts hormonogenesis and the gland is stimulated by TSH.[1][2]

The data (Fig. 2) show that before the goitrogenic challenge the labelling index and tritiated thymidine incorporation were very low, reconfirming that under normal conditions very few of the rat thyroid cells were in generative cycle. Within two days of

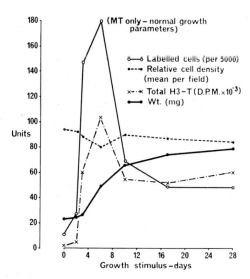

Fig. 2: Cell proliferation and DNA synthesis during goitrogenic challenge. No irradiation. Sequential measurements on rat thyroid before (0 days) and at intervals during continuous promotion of thyroid growth using 0.1 per cent methylthiouracil and low iodine diet for 28 days (goitrogenic challenge). Values are means from 5 animals per sacrifice. To be compared with Fig. 1.

administration of the goitrogen, however, both the labelling index and the tritiated thymidine incorporation increased markedly, then rose to high levels corresponding with the rapid growth of the thyroid till the eighth day and finally fell as growth slowed. Cell density dropped toward a small dip about day six followed by a slow fall, but the total decline was not substantial.

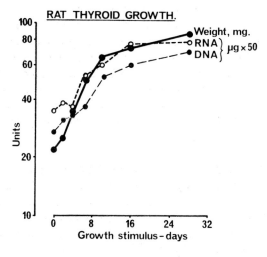

Fig. 3: Cell proliferation and RNA/DNA synthesis during goitrogenic challenge. No irradiation. Data are mean thyroid weight and chemical RNA/DNA—5 animals per group.

Experiment 3.

In this experiment thyroid weight and chemical DNA and RNA were measured before and during a goitrogenic challenge. Seven groups of five rats were used. It should be noted that the ordinate scale in Fig. 3 is logarithmic. The graphs showed that DNA and RNA both rose in parallel with thyroid weight, but not quite to the same extent.

Experiment 4.

Groups of three rats were sacrificed at times shown before and during goitrogen stimulation. Thyroid weight and stromal cells (including labelled) as a percentage of all cells were measured. Figure 4 shows that the proportion of stromal cells increased from 30% to 42%.

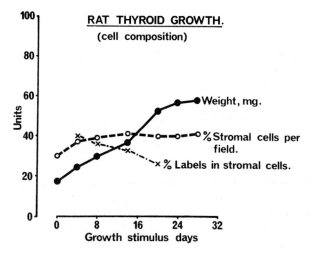

Fig. 4: Proliferation of stromal and follicular cells during goitrogenic challenge. No irradiation. Values are cells as percentage of total based on cell counts on 50 fields from 3 rats per sacrifice; per cent stromal cells is percentage of *all* cells; per cent labels in stromal cells is percentage of all *labelled* cells.

Experiment 5.

In this experiment groups of five rats were commenced at day 0 on the goitrogen which was continued for five days. From the 5th to the 12th day, however, the goitrogen was removed and the rats took water and standard diet only. On the 12th day the same goitrogen was recommenced and continued up to the 28th day. The points shown are the mean of thyroid total weight and the means of the weights of the left and right lobes respectively (L = left lobe and R = right lobe).

This experiment (Fig. 5) showed that a temporary cessation of the goitrogenic stimulus results in an immediate stop to the thyroid mass increase, but resumption of the stimulus after an interval of seven days produces a continuation of the mass increase, without an additional lag phase, proceeding as before. In this context, it has been said that if one thyroid lobe is removed at the time when weight has reached a maximum plateau (28 days) the remaining lobe does not double its weight but remains unchanged.[13]

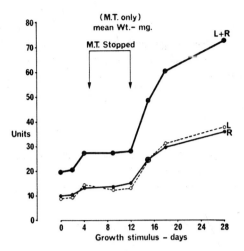

Fig. 5: Thyroid weight change pattern during temporary cessation of goitrogenic challenge. No irradiation. Goitrogen stopped from day 5 to day 12. Values are means from 5 animals per sacrifice.

Comment on goitrogen induced thyroid growth without irradiation (control).

In considering the actual sequence of events during the weight increase due to goitrogenic stimulation in the absence of irradiation a number of points of evidence must be linked. The initial low labelling index and low tritiated thymidine incorporation showed that cell turnover was very low before the goitrogen (Figs. 1 and 2). The high peaks in both of these indices during goitrogenic stimulation (Figs. 2 and 3) corresponded to the maximum rate of increase of both weight and chemical DNA. The low labelling index found as the growth curve flattens off showed that the number of cells synthesising DNA simultaneously was much reduced as the growth slowed. The near parallel increases in chemical DNA and RNA and mass (Fig. 3) show that on average the mean number of cells increases concomitantly with total cell mass and thyroid weight. Follicular cells do increase in size before division during goitrogenic hyperplasia but this change appears to be offset by loss of colloid. The net result is that cell density remains relatively unaltered during goitrogenic growth as shown in Fig. 2. These conclusions were also reached recently by Philp et al.,[5] and, in general, by Doniach.[14]

Goitrogen induced rat thyroid growth appears therefore to be a well regulated but special type of growth. The picture which can be drawn is of the majority of the thyroid cells responding, after a short lag phase during which hormone stores (colloid) are depleted, to the goitrogen by moving into active generative cycling. Synthesis of DNA and cell division appear to proceed in balance and maintain the normality of the cells and most of the organ growth (weight) is due to the increase in cell numbers. The number of cells synthesising DNA and dividing soon falls steeply, however, and the rate of growth slows to approach an apparent maximum asymptotically.

It might be postulated that this limitation of growth arises from extrathyroidal control, nutritional deficiencies in the goitrous state, or to factors intrinsic to the thyroid cells themselves. It is unlikely that extrathyroidal factors exist which are specific to only one thyroid lobe.[13] The evidence of experiment 4 showing a greater proportional rise in stromal cells (Fig. 4) is against the postulate of nutritional deficiencies. It would appear, therefore, that it is the thyroid cells which have an inherent limited divisional capacity seen as a ceiling or plateau limit to organ growth despite continued stimulation and adequate nutrition.

Although most of the weight increase is due to an increase in cell numbers, a significant part must arise from an increase in the average size of cells. The average contribution made by cell hypertrophy would appear to be the 20% which the total DNA, an index of cell number, failed, to rise proportionately to the total weight (Fig. 3). The proportional contribution to thyroid mass made by non-cellular structure other than colloid during goitrogenic growth is small.[1,15]

In the light of the comments made above concerning the cellular events occurring during the 28 day goitrogenic stimulation in the absence of irradiation the experiments using irradiation can now be presented. In these, latent damage was first produced by single doses of X-rays to normal rat thyroid then after an interval of 4 weeks the effects of the damage to the cell population was measured. It has previously been shown that latent radiation injury to thyroid cell reproductive integrity is permanent.[16]

Experiment 6: Effects of a goitrogenic challenge on irradiated thyroid.

Single X-ray doses of 100 rad, 500 rad and 1000 rad were given to groups of rats. All animals were commenced on the goitrogenic regime 4 weeks after irradiation. Subgroups of five animals from each X-ray dose group were sacrificed at the times indicated in Fig. 6, that is, just before and up to 28 days after the start of goitrogenic stimulation.

The effects of these various X-ray doses on the responses to the goitrogen as revealed by thyroid weight, cell density, total tritiated thymidine incorporation, labelling index and stromal cell percentage, are shown in Figs. 6 to 10 respectively. Each figure is discussed separately and includes for comparative reference the data from experiments 2, 3 and 4 (Figs. 1, 2, 3 and 4), in which non-irradiated thyroids were subjected to the same goitrogenic procedure (control).

Effect of X-irradiation on thyroid weight.

The greatest effect was seen after 1000 rad (Fig. 6). With this dose the thyroid

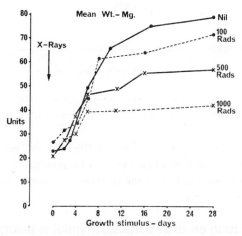

Fig. 6: Effects of X-irradiation on thyroid growth during goitrogenic growth stimulus, values are mean thyroids weights from 5 rats.

weight increased at a normal rate for nearly six days but then growth stopped. With 500 rad weight increased normally for six days and after six days growth was slow but not halted. 100 rad did not produce an effect significantly different from the unirradiated response, although a slight impairment of growth was discerned. Thus, for all three radiation doses the thyroid weight increased normally during early growth promotion but at about the 6th day of growth the weight curves in rats given 500 rad and 1000 rad X-rays plateaued and growth was thereafter much decreased, the degree of final impairment (weight at 28 days) being X-ray dose determined. It would appear, therefore, that the mode of action of X-irradiation in limiting the ultimate capacity of the thyroid to grow is linked to the later stages of growth.

Lack of effect of X-irradiation on cell density.

There was no significant difference (Fig. 7) between the sequential patterns of cell density in the irradiated thyroids and the controls, both showing a slight fall during growth promotion. This shows that, compared to normal, the ratio of cell number to cell size and non-cellular structure is unaffected by irradiation.

Effects of X-irradiation on stromal cell percentage.

The increase in stromal cells (percentage of total cells) during goitrogenic challenge was greater after 500 rad and 1000 rad than after 100 rad (Fig. 8) or no irradiation. This

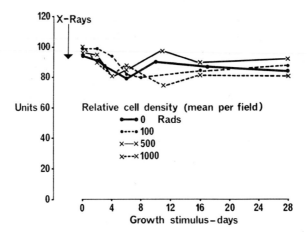

Fig. 7: Effects of X-irradiation on total thyroid cell density during goitrogenic growth stimulus, values are mean cell density in 50 fields from 5 thyroids.

effect may be due to differential X-ray impairing effects on cell proliferation within the two compartments, the stromal cells being more radio resistant than the follicular cells, or it might arise because stromal cells have the stimulating action of tissue injury added to that of the goitrogenic challenge.

Effects of X-irradiation on total tritiated thymidine incorporation.

The pattern of total gland tritiated thymidine ($T-^3H$) incorporation after a dose of 100 rad was not significantly different from that of the non-irradiated controls. (Fig. 9).

Measurement of Radiation Effects on Thyroid Cell DNA Synthesis using Tritiated Thymidine

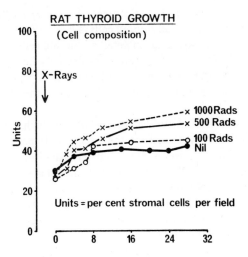

Fig. 8: Effects of X-irradiation on proliferation of stromal and follicular cells during goitrogenic growth stimulus. Values are per cent total cells based on cell counts in 50 fields from 5 rat thyroids.

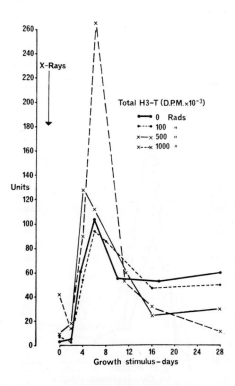

Fig. 9: Effects of X-irradiation on total thyroid incorporation of tritiated thymidine into DNA (total T-^{3}H) during goitrogenic growth stimulus. Values are d.p.m. from 5 pooled thyroids.

Since, however, total T–^{3}H uptakes, in the context of these investigations, are indices of average rates of DNA synthesis over all thyroid cells, a lack of effect of X-rays is not incompatible with some sublethal change in detail within the cells as will be discussed below. The measurements for 500 rad showed a considerably higher peak in DNA synthesis than normal between 4 and 8 days, the time when the most vigorous increase in cell numbers should have taken place (as judged from Fig. 6) and the 1000 rad peak was higher still.

These increased rates of DNA synthesis after both the higher X-ray doses must mean that the total amount of DNA synthesised in the intermediate period of goitrogenic growth is greater than normal. This is, however, approximately balanced by the lower rates and resulting lower total amounts of DNA synthesised at the last stages of growth (Fig. 9). Hence the total amounts of DNA synthesised do not appear to be greatly affected by X-irradiation but the aggregated synthesis seems to be performed in a shorter time. The increased total DNA synthesised in the first part of the goitrogenic response following 500 and 1000 rad could arise through three different mechanisms: more cells might be synthesising DNA at a normal rate or a normal number of cells could be synthesising DNA faster than normal, or fewer cells could be synthesising very much more DNA and rapidly. These possibilities were examined below in the light of the labelling indices data.

Effects of X-irradiation on labelling index.

In irradiated thyroid very large peaks in labelling index (labelled cells per 5000 cells) occurred at the same time as the unirradiated peak and after 1000 rad and 500 rad the labelling index was twice as high as in unirradiated gland (Fig. 10). The post-irradiation values reached 500 per 5000 cells compared to 250 for non-irradiated thyroid. These data are taken to mean that throughout the period of most vigorous cell DNA synthesis, twice

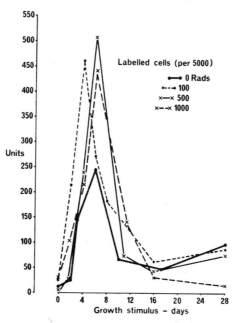

Fig. 10: Effects of X-irradiation on labelled thyroid cells (T–^{3}H autoradiographs) per 5000 total cells during goitrogenic growth stimulus. Values are from 5 thyroids.

as many irradiated cells, as control cells, were in active DNA synthesis, and since they synthesised about twice as much total DNA as non-irradiated gland (Fig. 9) the mean rates must have been approximately normal. After this phase of high labelling (Fig. 10) the labelling indices fell, and after 1000 rad the fall was to below normal levels. Following 100 rad the labelling indices reached as high an abnormal peak as after 500 rad or 1000 rad. This contrasts with the relative normality of the total tritiated thymidine uptake (Fig. 9). Thus after 100 rad, although total DNA synthesis rates were normal, the number of cells involved during rapid thyroid growth were about doubled. A possible explanation for this is either that repair of DNA is important or that a somewhat longer and slower S-phase within a cell cycle of unchanged length results from the moderate sublethal damage produced by the 100 rad; sublethal means here a subtle change in the cell generative cycle but not sufficient to alter viability or reproductive capacity. The subtle alterations in pattern of DNA synthesis after 100 rad (Figs. 9 and 10) which did not significantly impair goitrogenic growth (Fig. 6), contrasted with the effects of 500 and 1000 rad which were decisive.

Comment on radiation effects on goitrogen induced thyroid growth

The effects patterns which emerge with these high doses of X-irradiation and goitrogenic growth promotion over 28 days, are thus of an apparently normal initiation of growth and DNA synthesis in the thyroid but the number of cells which are ultimately able to divide is decreased as a result of the irradiation. It would seem that the large number of cells which do not pass into normal mitosis can make DNA. As a consequence, although growth is impaired or arrested at the phase of maximum potential increase (Fig. 6), the total uptake of tritiated thymidine and the proportion of cells labelled is higher than normal (Figs. 9 and 10). This interpretation is also consistent with the quick fall in total DNA synthesis, this probably showing that the number of cells involved after thyroid growth has ceased prematurely falls off sharply.

Other investigators[17] have incidentally noted increased uptake of tritiated thymidine after irradiating rat thyroid and giving iodine deficiency as a goitrogenic stimulus. Like them we think the increased cell DNA synthesis is the radio-biochemical counterpart of the large abnormal nuclei noted in morphological preparations of human thyroid following therapeutic doses of irradiation.[18,19] These data show a high level of apparent abortive DNA synthesis in irradiated cells when they are stimulated into generative cycling. This could also be the radiobiochemical equivalent of chromosome breaks and their repair noted by other investigators examining mammalian thyroid irradiated *in vivo*.[20,21]

Increased DNA synthesis in other mammalian cells after their irradiation *in vivo* and during induced cycling has also been recently reported by Smets,[22] and Watanabe and Okada.[23]

Subsidiary experiment(s) 7.

In order to establish the radiobiological significance of the above observations and exclude artefacts, a number of subsidiary validation experiments were carried out. These will be reported only briefly and their relevance to present investigations summarised.

The possibility was considered that the effects of the goitrogenic challenge on tritiated thymidine incorporation and labelling indices might have been an artefact produced by some peculiarity of the drug methylthiouracil. Such a drug effect might have been produced, for example, by a change in the availability time of the tritiated thymidine through a systemic effect of methylthiouracil on body or intrathyroidal thymidine distribution and disposal. This possibility was excluded with three experiments.

In mice given tritiated thymidine intraperitoneally the retro-orbital blood tritium radioactivity time curve determined by the method of Hansen and Bush[9] was the same as

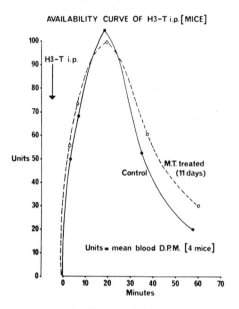

Fig. 11: Mouse blood tritiated thymidine (d.p.m.) curves after a single intraperitoneal dose (50 μCi). Comparison of curve in normals and methylthiouracil treated mice, values are means from same 4 mice.

that of untreated control mice (Fig.11). In rats, the sequential changes in thyroid DNA incorporation of tritiated thymidine measured up to 6 hours after a single administration of tritiated thymidine to methylthiouracil primed animals showed increasing uptake till a plateau commenced at 4 hours after thymidine administration and the thyroid uptake-time curve co-related with the heart blood tritiated thymidine-time curve (Fig. 12). Another experiment demonstrated that within the limits of specific activity used (17.0 to 28.0 Ci/mM), the rat thyroid incorporation of the tritiated thymidine showed little variation and this applied whether rapid thyroid growth was induced by absolute iodine deficiency or by methylthiouracil (Fig. 13). This appeared to exclude a possible pharmacological effect of the drug on the thyroid itself. These studies appear to prove that the changes in tritiated thymidine incorporation and labelling indices produced in the thyroid during the goitrogenic regime were not artefacts but reflected the real behaviour of the DNA synthetic processes in normal cells undergoing proliferation. There is no reason to suspect that equal validation applied to X-irradiated thyroid but controls in all these above respects were not practical.

DISCUSSION AND CONCLUSIONS

These studies demonstrate that unirradiated cells in *normal adult* rat thyroid synthesise DNA and divide but at very low rates. The failure of a single dose of 500 rad of X-rays

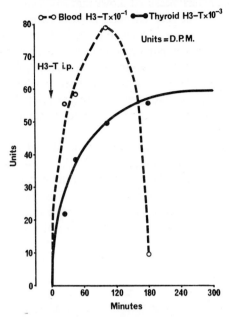

Fig. 12: Rat thyroid cell DNA tritiated thymidine (d.p.m.) and heart blood tritiated thymidine (d.p.m.) sequentially after a single intraperitoneal dose (0.5 µCi/gram). Values are means from 2 different pairs.

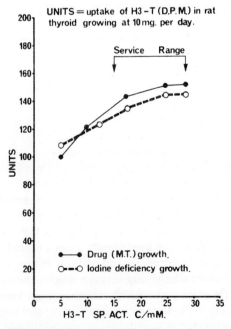

Fig. 13: Rat thyroid cell DNA tritriated thymidine (d.p.m.), 4 hours after single intraperitoneal dose (0.5 µCi/gram). Effects of varying the Sp. Activity; iodine deficiency growth compared with methyl thiouracil growth. Values are means from different pairs.

to perturb any of the parameters measured shows that any damage produced is not demonstrable using gland weight, cell density, chemical DNA or tritiated thymidine incorporation as the indices (Fig. 1). A single dose of 500 rad is a large dose by radiobiological standards producing gross disturbances in cell populations in active proliferation.[24] Since there is no migration of cells to and from the thyroid and the thyroid cell population is mainly in G_1 or G_0 damage produced by 500 rad or more is latent. The other studies demonstrate this, since by promoting cell proliferation with a goitrogenic challenge marked changes are seen when non-irradiated and irradiated thyroid is used for comparison (Figs. 2, 3 and 4 compared to Figs. 6 to 10 inclusive).

The effects of the goitrogenic challenge in non-irradiated animals appears mainly on nucleic acid synthesis, cell division and organ growth, the rates of which, after a short lag, rise to a maximum and decline quickly indicating that the system is self-regulated (Figs. 2, 3 and 4). The weight increase, due mostly to an increase in the number of cells, for example, increases quickly to approach asymptotically to a maximum weight. This restriction to a maximum weight appears to be due to a well regulated ceiling, the number of cell divisions being estimated as an average of not more than one or two per cell. The limitation of division appears to be intrinsic in the adult thyroid cells themselves, a conclusion recently reached by Sheline[25] too, but the current studies cannot show whether any of the cells have different intrinsic divisional capacities. The cell population has therefore to be considered as one population but with follicular or stromal cell subcompartments.

The irradiation effects (Figs. 6 to 10 inclusive) brought out in thyroid subjected to goitrogen stimulation were X-ray dose-dependent. However, only net thyroid growth as determined by weight at the end of a 28 day goitrogenic regime, correlated directly with X-ray dose (Fig. 6).

In terms of the mode of action of X-rays the overall results are interpreted as showing that the principal effect of latent radiation damage in the thyroid is to reduce the proportion of cells which are able to divide when called upon to do so and the number of divisions they can complete but some effect on capillary integrity cannot be excluded.[26] Although the rates of DNA synthesis are severely affected by X-rays (Figs. 9 and 10) they do not alone form a simple quantitative index of radiation damage. Continued DNA synthesis without cell division and tissue growth makes interpretation of the radiation effect, if measured by DNA synthesis alone, very difficult.

It would thus appear that the most appropriate simple index of crude cell survival after thyroid irradiation is the net impairment of thyroid growth to the whole 28 day goitrogenic procedure; but net weight is final thyroid weight minus initial thyroid weight. Weight increase not due to cell proliferation but to an increase in cell size or in the non-cellular structures in the thyroid must be considered before the system is adopted for quantitative cell survival studies. Cell hypertrophy is not a special feature of goitrogen stimulated rat thyroid since cells do enlarge and multiply in sequence and simultaneously[27] so that there is a near constant relationship between weight, chemical DNA and RNA before and throughout the goitrogenic growth challenge (Figs. 1 and 3), The relatively steady index of cell density in the same conditions (Fig. 2) also suggests that increased cell number is the dominant change in goitrogen promoted thyroid growth. We estimate that the fraction of net thyroid growth not due to cell division is about 10%.[28]

Thus, when the rat thyroid, stimulated by a goitrogenic challenge, is to be employed as an index of residual thyroid cell survival not only must the goitrogenic regime be continued for at least 28 days but the growth should be expressed as net growth (final gland weight minus initial gland weight) and in addition a correction should be made for growth not due to cell division. The latter, about 10% in value, should be obtained in each radiation experiment; this may be simply measured as the net growth which is constantly retained after doses of irradiation in the very high radiobiological range (e.g. more than a single X-ray dose of 1200 rad).

In conclusion, the rat thyroid may be employed as a model with which to study thyroid cell survival *in vivo* after irradiations. The special features of the model are that it is one of a highly differentiated tissue whose cells cannot divide indefinitely but perhaps only once or twice. It has, however, the advantage of relative simplicity and the results of experiments are relevant to the biological consequences and the therapeutic effects of ionising irradiations on human thyroid. They are also likely to be relevant to clarifying radiobiological effects on organised differentiated tissues in general.

For example, the studies are relevant to the observation that irradiation and goitrogenesis are carcinogenic in rodents.[3] The current studies explain why irradiation of the normal foetal infant and child thyroid is much more likely to lead to subsequent hypothyroidism or thyroid cancer than irradiation of the adult organ.[9,30,31] When the young thyroid which has not completed its growth, is irradiated presumably further normal physiological growth is impaired and abnormal cells are left; the impairment of physiological growth being equivalent to the impairment of goitrogenic growth in the rat and studies above. Radiation impairment of thyroid cell reproductive integrity is also likely to explain why so many patients eventually become hypothyroid after iodine-131 therapy for thyrotoxicosis.[32] Finally, studies of the type described here may well allow investigations to prove or disprove some theories about the fundamental cell changes of radiation carcinogeneus.[33]

ACKNOWLEDGEMENTS

This work is supported by grants from the Wellcome Trust and The Cancer Research Campaign. Figs. 1 to 10 are reproduced by kind permission of the editor of the International Journal of Radiation Biology **16**, 211 (1969) and my colleagues, Drs. J. F. B. Smith, W. P. Duguid, C. J. Foster and J. S. Orr, respectively.

REFERENCES

1. J. E. Santler, *J. Endocrinol.* **15**, 151 (1957).
2. C. W. Gilbert and L. G. Lajtha in *Cellular Radiation Biology,* Williams and Wilkins Co, Baltimore, 1965, p. 474.
3. I. Doniach, *Brit. Med. Bull.* **14**, 181 (1958).
4. J. M. Gibson and I. Doniach, *Brit. J. Cancer* **21**, 524 (1967).
5. J. R. Philp, J. Crooks, A. G. MacGregor and J. A. R. McIntosh, *Brit. J. Cancer* **23**, 515 (1969).
6. J. Crooks, W. R. Greig, A. G. MacGregor and J. A. R. McIntosh, *Brit. J. Radiol.* **37**, 380 (1964).
7. J. D. Begg, E. M. McGirr and H. N. Munro, *Endocrinology* **76**, 171 (1965).
8. J. F. Moorehead and W. McFarland, *Nature* **10**, 1157 (1966).
9. D. L. Hansen and E. T. Bush, *Anal. Biochem.* **18**, 320 (1967).
10. D. R. White, *Intern. J. Appl. Radiation and Isotopes* **19**, 49 (1968).
11. B. M. Kopriwa and C. P. LeBlond, *J. Histochem. Cytochem.* **10**, 269 (1962).

12 A. Brodish, *Yale J. Biol. Med.* **41**, 143 (1968).
13 I. Doniach, in S. Young and D. R. Inman (Eds) *Thyroid Neoplasia,* Academic Press, New York, 1968, p. 265.
14 I. Doniach, *Brit. Med. Bull.* **16**, 99 (1960).
15 S. Y. Chow and D. M. Woodbury, *Endocrinology* **77**, 825 (1965).
16 W. R. Greig, J. A. Boyle, W. W. Buchanan and S. Fulton, *J. Clin. Endocrinol. Metab.* **25**, 1009 (1965).
17 B. M. Dobyns, A. Rudd and M. A. Sanders, *Endocrinology* **81**, 1 (1967).
18 B. M. Dobyns and I. Didtschenko, *J. Clin. Endocrinol. Metab.* **21**, 699 (1961).
19 B. M. Dobyns and L. R. Robinson, *Clin. Endocrinol.* **28**, 875 (1968).
20 W. Moore Jr and M. Colvin, *Int. J. Radiation Biol.* **14**, 161 (1968).
21 J. W. Speight, W. I. Baba and G. M. Wilson, *J. Endocrinol.* **42**, 267 (1968).
22 L. A. Smets, *Int. J. Radiation Biol.* **14**, 585 (1968).
23 L. Watanabe and St. Okada, *Radiation Res.* **35**, 202 (1968).
24 T. Alper, in T. J. Deeley and C. A. P. Wood (Eds) *Modern Trends in Radiotherapy,* Butterworth and Co, London, 1967.
25 G. E. Sheline, *Cell and Tissue Kinetics* **2**, 123 (1969).
26 S. P. Stearner and E. J. B. Christian, *Radiation Res.* **34**, 138 (1968).
27 H. A. Johnson, *Am. J. Pathol.* **57**, 1 (1969).
28 W. R. Greig, J. F. B. Smith, J. S. Orr and C. Foster, *Brit. J. Radiol.* **43**, 542 (1970).
29 J. Robbins, J. E. Rall and R. A. Conrad, *Ann. Internal Med.* **66**, 1214 (1967).
30 L. H. Hempelmann, *Science* **160**, 159 (1968).
31 G. W. Dolphin, *Health Phys.* **15**, 219 (1968).
32 W. R. Greig, *J. Clin. Endocrinol. Metab.* **25**, 1411 (1965).
33 W. V. Mayneord, *Brit. J. Radiol.* **41**, 241 (1968).

DISCUSSION

L. Schutte: Concerning discrimination between iodine-125 and tritium activity, did you consider combustion of your samples and subsequent chemical separation of iodine (e.g. as iodide) and water, after which you can count them separately?

W. R. Greig: Yes, this is one approach that we have in mind.

C. P. Summers: In practice in a liquid scintillation counter, it would not be possible to separate the β or electron emission with the purpose of distinguishing between iodine-125 and tritium spectra because the gamma emissions from iodine-125 would produce Compton electron, i.e. further β particles which would completely mask the β spectra. The best way of doing this measurement is to measure iodine-125 in a gamma counter and then iodine-125 and tritium in a liquid scintillation counter, having previously determined iodine-125-iodine-125/tritium counting ratio.

W. R. Greig: This is a valid point. I think the problem of counting small amounts of tritium in the presence of relatively large quantities of iodine-125 is not likely to be solved by the above approach; I suppose, however, that with careful calibration and appropriate controls the limit of this method could be defined, and should be.

B. Legg: Concerning the question of tritium and iodine-125 counting by liquid scintillation, surely a window can be selected above the tritium window to count the iodine-125, the overlap in the tritium window determined, and the tritium count found by difference. Provided the tritium: iodine-125 ratio is above a certain value the result should be accurate. Alternatively, the sample can be counted and recounted after, say, one week when an appreciable amount of iodine-125 will have decayed. The iodine-125 can be calculated from the data and the tritium is again obtained by difference.

W. R. Greig: Certainly one would intuitively think that by multiple channel counting the tritium count could be distinguished from the iodine-125 counts but no-one has formally conducted this study and it is clear from this discussion that a project lies in this area. Recounting after a differential decay is difficult because during the latter the liquid scintillation system in the vial deteriorates.

B. W. Fox: Have you compared the tritiated thymidine incorporation following the damage from iodine-131 and iodine-125?

W. R. Greig: Not directly, but I have carried out autoradiographs of the labelled cell nuclei. Iodine-125 leaves more labelled cells than iodine-131 in equivalent rad doses (see W. R. Greig, PhD Thesis, University of Glasgow, 1970).

J. A. B. Gibson: What is the cell cycle time following 'treatment'? What evidence do you have vis a vis repair and an increase in the S phase which would increase the cycle time? It would obviously be useful to look at synchronised cells *in vitro* to see if S is increased or if there is take up during G1 or G2 indicative of repair.

W. R. Greig: I do not know the answers to these questions but I am pursuing the work using rat thyroid irradiated *in vivo,* cultured and labelled *in vitro.* Your comments are helpful.

K. R. Harrap: (i) Have you succeeded in demonstrating a differential effect against thyroxine synthesis in rat thyroid, comparing iodine-125 and iodine-131? (ii) Would you expect to eliminate thyroxine synthesis without impairing transcription of genetic information via prior irradiation of DNA?

W. R. Greig: (i) Using a variety of techniques I have attempted this and so far do not find that iodine-125 has a proportionally greater effect than iodine-131 on hormongenesis in rats but it might have in human thyrotoxicosis. (ii) On the basis of our current knowledge we might expect to impair the final stages of thyroxine synthesis and further studies are planned in this area.

Chapter 12

Quantitative Studies of Enzymes and Drug Actions in Cells and Microslices using ^{14}C-labelled Substrates

G. A. Buckley, C. E. Heading and J. Heaton

Department of Pharmacology and General Therapeutics,
University of Liverpool, England

INTRODUCTION

Reed[1] has reviewed the methodology of radiotracer enzyme assays, illustrating the advantages and disadvantages and giving a list of more than 250 enzyme assay methods using radioisotopes. Many of these methods have not been used or adapted for the microscale but the application and value of isotopic methods to cytochemistry have been reviewed by McCaman.[2,3]

During the past few years interest in our laboratory has centred on the relationship between enzyme levels and the organisation and function of the nervous system. Originally we used colorimetric methods with a 'sensitivity limit' of approximately 10^{-10} moles.[4] For this purpose sensitivity limit means the minimum detectable amount of product from the enzyme reaction. Microgasometry, fluorimetry and radiochemistry were the possibilities for increased sensitivity. We chose to use radiochemical methods because of their simplicity and versatility.

METHODS

The work has been concerned with the enzymes cholinesterase and cholineacetylase. To estimate these enzymes it is necessary to separate the labelled products, acetate and acetylcholine, from the labelled substrates acetylcholine and acetyl-CoA, respectively. We consider that for a micromethod it is essential to be able to obtain the product of the reaction in one step, e.g. extraction or precipitation.

The method for cholinesterase is illustrated thus:

$$(^{14}C\text{-acetyl}) \text{ choline} \rightarrow {}^{14}C\text{-acetate}$$

On addition of acid to pH 1 the ^{14}C acetate is extracted into toluene/isoamyl alcohol leaving the ^{14}C acetylcholine in the aqueous phase.

The method for cholineacetylase is illustrated thus:

$$(^{14}C\text{-acetyl})\text{-CoA} + \text{choline} \rightarrow (^{14}C\text{-acetyl}) \text{ choline}.$$

The ^{14}C-acetylcholine is extracted using Kalignost dissolved in 3-heptanone, the ^{14}C-acetylCoA remaining in the aqueous phase.

The volume of the initial reaction mixture is usually about 1 μl, and the volume of organic solvent containing the labelled product is usually about 50 μl. An important consideration in assessing the value of these methods is that, because of the small amounts of material used, the costs are low enough to be discounted.

RESULTS
Cholinesterase at the Myoneural Junction

Most of the information about cholinesterase at the myoneural junction has come from histochemical staining techniques. Whilst giving good information about the localisation of the enzyme these methods yield little or no quantitative data. Colorimetric methods are not sensitive enough to tackle the problem and microgasometric methods are too slow and cumbersome to give sufficient data for statistical analysis. The radiotracer method has been used with much success to answer several important questions.

Buckley and Heaton[5] showed that the myoneural junctions of extraocular muscles have a wide range of cholinesterase activities. They also showed that myoneural junctions from singly innervated fibres have higher cholinesterase activities than myoneural junctions from multiply innervated fibres. Subsequently we have shown that myoneural junctions from twitch fibres have higher cholinesterase activity than tonic fibres in muscles from frogs, chickens, rats, cats, rabbits and man. The level of cholinesterase does not appear to be simply related to speed of contraction however, since there appears to be little or no difference between the cholinesterase activities of myoneural junctions from slow as opposed to fast twitch muscles.

A further important series of data that the radiotracer method has yielded is the changes of cholinesterase activity at the myoneural junctions during postnatal growth. Histochemical evidence, suggested that the level of cholinesterase in rat myoneural junctions does not change appreciably after the age of three weeks. We have shown, however, that whilst there is a very rapid increase during the first 3 postnatal weeks there is also a large increase after this time. The median cholinesterase activity of myoneural junctions from rat gastrocnemius at one week is 18 p.mol/junction/hr, at 3 weeks is 105 p.mol/junction/hr and at 52 weeks 350 p.mol/junction/hr. There is a similar picture in a twitch muscle of the chick but in a parallel tonic muscle there is only a slight increase in cholinesterase activity at the myoneural junction.

Finally, analysis of the kinetics of hydrolysis of ^{14}C-acetylcholine by dissected myoneural junctions from rat, frog and chicks has produced evidence suggesting that there are species differences in the organisation of the enzyme at this site. Purified cholinesterase has an optimum substrate concentration of about 3mM acetylcholine, higher concentrations causing substrate inhibition.

This type of kinetics was also found with myoneural junctions from chick muscles.

In the case of rats and frogs, however, the optimum substrate concentration for dissected myoneural junctions was higher than 20 mM but after homogenisation the optimum was 3 to 5 mM.

Cholinesterase and Cholineacetylase in single cells and micro-slices of sympathetic ganglia from rats and cats.

Buckley, et al.[6, 7] modified an existing method to enable them to measure the amount of cholineacetylase in single cells of cat sympathetic ganglia. They used the method to show that only about 13% of cells in the L7 ganglion of the cat had cholineacetylase activity higher than the sensitivity limit of the method (10^{-12}-10^{-13} Moles Ach). They suggested that these cells were cholinergic and the remaining cells were adrenergic.

Since then the method has been used in an investigation of rat sympathetic ganglia. In this study freeze-dried sections of ganglion tissue were weighed (approx. 1 μg) and cholinesterase and cholineacetylase measured. The data revealed that the various ganglia studied differ markedly in the absolute and relative contents of the two enzymes.[8] Alternative methods of measuring the amount of acetylcholine produced involve colorimetry or bioassay. The latter is more sensitive than colorimetry but is less sensitive and considerably more time-consuming than the radiometric method.

CONCLUSIONS

Radiotracer methods for the assay of cholinesterase and cholineacetylase have been used to study various physiological and pharmacological problems.

The information so obtained is ample evidence that radiotracer methods of enzyme assay adapted to a microscale provide a powerful analytical tool for the biologist. The techniques are not difficult to acquire and are simpler than many methods of comparable sensitivity. The cost/effectiveness ratio for the methods is comparatively low.

REFERENCES

1. D. J. Reed, *Advan. Tracer Methodol.* **4**, 145 (1968).
2. R. E. McCaman, *Isotopes and Radiation Technol.* **3**, 328 (1966).
3. R. E. McCaman, *Advan. Tracer Methodol.* **4**, 187 (1968).
4. G. A. Buckley and P. T. Nowell, *J. Pharm. Pharmacol.* **18**, Suppl. 146S (1966).
5. G. A. Buckley and J. Heaton, *J. Physiol. (London)* **199**, 743 (1968).
6. G. A. Buckley, S. Consolo, E. Giacobini and R. E. McCaman, *Acta Physiol. Scand.* **71**, 341 (1967).
7. G. A. Buckley, S. Consolo, E. Giacobini and F. Sjoqvist, *Acta Physiol. Scand.* **71**, 348 (1967).
8. C. E. Heading, *Brit. J. Pharmacol.* **37**, 533 P (1969).

DISCUSSION

P. J. England: With regard to the precision of your estimates of cholineacetylase activity, is there any chance of micro-reversibility of the enzyme, i.e. are counts being incorporated into acetylcholine without net flow of material? This would have the effect of giving higher estimates of the rate of the reaction than are actually occurring.

G. A. Buckley: The equilibrium for the cholineacetylase reaction is very much in favour of acetylcholine production so that, under our conditions, the reverse reaction is negligible.

Chapter 13

Critical Remarks about Current Trends in Liquid Scintillation Counting

D. A. Kalbhen

Institute of Pharmacology, University of Bonn, W. Germany

In the development of instruments and techniques in liquid scintillation counting within the past 15 years one can observe that this very useful counting technique becomes more and more complex and may sometimes be quite confusing for beginners. The instrumentation has become more sensitive and delicate (although more automated) and the number of sample preparation techniques, solvents, solutes and solubilisers has increased enormously. There are about 100 to 200 different scintillation solutions which are described in the literature and nearly every researcher uses his own 'modified' version. For each variation in solvent, solute, and sample preparation technique, it is necessary to recalibrate the liquid scintillation counter to optimal counting conditions. It has become quite impossible to quantitatively compare counting results from one laboratory with those from another.

This pattern of variation and alteration in counting conditions is mainly due to the fact that tracer samples, especially those of biological origin, have so many different physico-chemical properties and need therefore special methods of sample preparation. The principal factors which influence counting conditions and efficiency are the following:

1. Background (liquid scintillation counters cannot automatically handle background values in quenched samples).
2. Phosphorescence.
3. Chemiluminescence.
4. Properties of counting vials.
5. Quenching: colour and/or chemical.
6. Instability of samples: phase-separation, precipitation, chemical alteration.
7. Instability of instrumentation.
8. Application of non-suitable quench-correction-curves for the calculation of d.p.m.

To cope with these various factors, and to handle all kinds of tracer samples, the manufacturers of liquid scintillation counters have developed more and more sophisticated instruments with higher efficiencies and sensitivities, and with quite elegant data calculating systems (the latter reach from mechanical calculators to off and on-line computers). Besides being very expensive, the running of modern liquid scintillation counters with inbuilt or external computers requires intelligence, understanding and working experience.

In the course of this symposium and in regard of the current status of liquid scintillation techniques it becomes quite clear that the development of instruments and counting techniques is oriented only to the task of handling the many different tracer samples. Since about 90% of all samples counted by the liquid scintillation technique contain carbon-14 or tritium from biological specimens or biochemical experiments, it appears to be reasonable to standardise the sample preparation techniques in order to obtain a more or less uniform counting sample.

I would like to suggest that all users of the liquid scintillation technique, as well as all manufacturers of liquid scintillation counters, put more effort into the elaboration of standardised sample preparation methods and instruments, which (for instance by oxygen combustion) convert all biological materials into the same combustion products. Prototypes of this kind of instrument are already made by Packard Instruments. Although the final stage of development in this direction is not yet reached, these methods and instruments may become most useful and versatile. It is quite possible that an automatic sample preparation instrument can be developed which will handle not only carbon-14 and tritium samples but also sulphur-35, phosphorus-32, calcium-45 and others. In combination with only one scintillation solution automatically and uniformly prepared tracer samples can be counted in standardised liquid scintillation counters without the complicated data calculating processes which are necessary today.

DISCUSSION

L. Schutte: What is the effect of halogens, or other elements, on quenching or oxidation yield in the general combustion procedure? (e.g. CCl_4 cannot be oxidised).

D. A. Kalbhen: Only large amounts of halogens and other elements with quenching properties will interfere with counting efficiency. In biological specimens variations are rarely too broad (regarding e.g. halogen content, etc.). The counting of CCl_4 will always be problematic but calculable.

D. S. Glass: I would like to take issue with the point that a standard method should be used for all liquid scintillation counting to accommodate the large majority of samples counted which are of biological origin. To reduce liquid scintillation counters to a basic instrument for measuring a standard scintillation mixture ignores the important property of the instrument, that it can be an extremely accurate and rapid analytical instrument for concentration measurements. This property apparently is not fully exploited, and instrument manufacturers should remember this when turning their attention to their biochemica users.

D. A. Kalbhen: I also support the idea to use the counters for other analytical (photometric) purposes, as I mentioned in my paper about chemi- and bio-luminescence in liquid scintillation counting. But for this the instrumentation can be quite simple and reduced (e.g. single photomultiplier).

C. P. Bond: We have measured quite a lot of sulphur-containing compounds by 'oxygen-bag' combustion and never noticed any extra quenching problems over those of carbon-containing compounds.

D. A. Kalbhen: I agree with your remarks. We, too, had no problems counting sulphur-containing samples (using the wet-combustion technique by Mahin and Lofberg [see chapter 1, ref. 18]). $H_2{}^{35}SO_3$ or $H_2{}^{35}SO_4$ formed by oxidation can easily be counted by liquid scintillation with minimum quench observed.

Index

Acetyl choline 123, 124
Alcohols. 2, 10
α-particles (emitters) 27, 36, 54
α-spectrum 46
Aluminium stearate 97
Americium-241 27
Amino acids. 2, 10
p-Aminosalicylate. 12
Anode characteristics. 17, 19, 50
Anthracene. 31–34
Armeen L-11 98
ATP (adenosine triphosphate) 10, 11
Auger electrons 49, 51, 53
Autoradiographs. 106, 114, 121
Barium carbonate. 97–99, 101, 103
BBOT (2,5-bis-(5'-t-butyl-
    benzoxazolyl (2'))-thiophene) 2
Beckmann LS-100 spectrometer. 70
β-particles (emitters). 23–36, 37, 120
BHT (di-t-butyl-4-hydroxytoluene) 8
Biological tissue 2, 6
Bioluminescence assay 11
bis-MSB (p-bis-(o-methylstyryl) benzene). 2
Bleaching agents. 3, 5
Blood. 6, 103, 116
Bosons 40
Butyl-PBD (2-(4'-t-butylphenyl)
    -5-(4"-biphenylyl)-1,3,4-oxadiazole). . .2, 101, 102
Cab-O-Sil 97–101, 103
Caesium-137. . . . . .70, 99
Calcium-45. . . . . .1, 99, 128
Carbon-14 . . . . .1, 15, 20–22, 69, 71, 74, 77, 79–83,
                     87, 100, 102, 123–125, 127
Carbon tetrachloride 81, 128
Carrier 51, 52
Cary spectrophotometer. .26
Cell(s). 2, 105–121, 123–127
    counts. 109
    proliferation 108, 109
Čerenkov effect 36
Channels ratio technique 12, 57, 58, 62, 76, 77,
                           81–87, 91, 92
Chemiluminescence 1–13, 127
    assay 10
    decay time. 13
Chloroform 76
Cholineacetylase. 123–125
Cholinesterase 123–125
Chromium-51 . . . . .49, 52

Cobalt-57 53
Cocktails 3, 7, 9, 15, 34, 49, 54, 69, 70, 80, 101,
                             102, 106, 127
Coincidence technique. 13, 21, 50, 51, 53
Combustion technique. . . . 9, 102, 103, 120, 128, 129
Compton electrons. 120
Computer(s) (methods, on-line) 55–67, 79–95
    desk-top 69–78, 81, 79–95
Counting conditions. 80
    efficiency 17, 19, 51–53, 55, 56, 70, 75–77,
                       81–88, 101–103, 127
    errors 52, 60, 61, 64, 67, 70
    equipment. 15, 70, 71, 76, 79
Crystal scintillation counter. 49
Current trends 127
Decay times 25, 44
Diehl transmatic calculator. 81, 88, 94
Dimers 47
1-Dimethyl-aminonaphthalene-7-
    sodium-sulphonate 28
Dimethyl-POPOP 2
Dioxane 2, 3, 5, 12, 35, 70, 76
    scintillators .80, 102, 104
Discrimination characteristics. 17, 18, 50
DNA (deoxyribo nucleic acid) 105–121
Double label counting 54, 74, 90, 120, 121
Double ratio technique .12, 92
DPS (p, p'-diphenylstilbene) 44
Drugs 123
Electron capture nuclides. 49–54
Elliot 903 computer. 81, 91–95
Emission spectra. 23–36, 47
Enzymes. 11, 123
2-^{14}C-ethan-1-ol-2-amine hydrochloride 70
External standardisation (AES) 56–67, 70, 73,
                        77, 79, 81, 87, 91, 92
Figure of merit. 54
Fluorescence lifetime 13
FMN (flavin mononucleotide) 11
Fortran. 89, 91
Ganglia. 125
γ-rays 27
Gels (gelifying agent) 97–104
Glucose 10
Glycols. 2
Goitrogenic challenge. 105–121
    regime 105–121
    stimulus 105–121
Helium (liquid). 37–47

^{14}C-n-Hexadecane 15, 17, 18, 69
^{3}H-n-Hexadecane 69
HYAMINE-10X 2, 11
Hylene TM-65 98
IBM 1130 computer 81, 84, 91, 93, 94
IME 86S desk-top calculator 72
Imperial Data-log typewriter 72
Insta-gel 8
Iodine-125 . . . . 49, 54, 120, 121
Iodine-131 . . . . 99, 119, 121
Ionisation chambers 49
Iridium-192 28
Iron-55 . . . . 49, 50, 52, 53, 99
Iron-59 99
λ-point 37–41
Least squares method 72–75
Light pipe 21, 44
Luminescence spectra 23–35
Luminol reaction 10
Machine errors 60
Manganese-54 . . . . 49, 51–53
Methyl-^{3}H-choline chloride 70
Methyl-thiouracil 107, 108, 115, 116
Microslices 123
Monroe digital printer 79
Muldivo 72, 95
Muscle fibres 124
Myoneural junction 124, 125
NADH (reduced nicotinamide adenine
  dinucleotide) 11
Naphthalene 2, 5, 35, 70, 80
Neutron beam polarisation 43–47
NCS 2–11, 106
Nitric acid 8
Nitroaniline 10
Nuclear Chicago spectrometer 3, 76
Nucleic acids 2, 106
Olivetti desk-top calculator 81, 84, 88, 90, 93, 94
Organic peroxides 5, 6, 10, 13
Organophosphorus compounds 10
Osmium lamp 13
Oximes 10
Packard 3, 79, 80, 129
Parallel counting system 17–21, 50, 51, 53
PBD (2-(4-biphenyl)-5-phenyl-
  1,3,4-oxadiazole) 2, 15, 18, 22, 49, 51, 52
PBO (2-phenyl-5(4-biphenyl)-1,3-
  oxazole) 34
Perchloric acid 8, 9, 106
Phenoles 10
Phenylethylamine 2
Phonons 39–41

Phospholipids 70, 74
Phosphorescence 21, 127
Phosphorus-32 . . . . 20, 128
Photocathode 45
Photoemission 37
Photomultipliers 15, 20, 26, 29, 44–46, 60, 128
  spectral sensitivity 30
Photons 49, 51–53
Plasma 102
Plastic scintillators 53
Plates (thin-layer) 99
Polonium-210 46
Poly-Gel-B 98–103
Poly-Gel-B13 . . . . 102, 104
Poly-Gel-11 104
Polyglycol ethers 2, 8
Polynomials 59, 66, 67, 72, 81, 88, 90
Polysaccharides 2
POPOP (1,4-bis-(5-phenyloxazol-2-yl)benzene) 2,
  22, 33, 44, 49, 54, 70, 80, 97, 98, 101, 102, 106
Potassium hydroxide 1–11
PPO (2,5-diphenyloxazole) 2, 15, 18, 22, 35, 49,
  54, 70, 80, 97, 98, 101, 102, 106
PRIMENE 81-R 2
Programming 55, 72, 77, 87
Promethium-147 28
Proteins 2
Punch cards 89, 91
  tape 69, 78, 89, 91, 93
Pyruvic acid 12
p-Quaterphenyl 35
Quenching 2, 6, 9, 10, 21, 25, 26, 50, 51, 53,
  61, 81, 100, 127, 128
Quench correction (curves) 55–67, 69–73, 76,
  77, 79, 127
Radiation effects 105–121
Radium-226 79
Rat gastrocnemius 124
  liver powder 99, 102
  thyroid 105–121
Reflector paint 44, 46
Rhodamine B 28
RNA (ribo nucleic acid) 105–121
Rotons 39–41
Secondary pulses 19, 20
Scintillation efficiency 21, 23, 54
Singlet state 26
Sodium hydroxide 1–11
Solvents 2
Solubilisers 2, 9
Soluene-100 2–11
Solutes 2

Stannous chloride	13
Statistical analysis	63
Strontium-90	99
Sulphur-35	1, 128, 129
Suspension counting	101
p-Terphenyl	34
Thixcin (ricinoleic acid)	97
Three-point method	72, 74, 75
Thymidine-6-^{3}H	105, 107, 111
Thyrotoxicosis	119, 121
Time distribution	20
trends in counting	64
Toluene scintillators	2, 3, 8, 9, 13, 15, 31–35, 49, 51, 52, 54, 70, 76, 78, 80, 97, 101, 102, 106
Triplet state	21, 26
Tritium	1, 7, 22, 54, 69, 71, 74, 75, 79–81, 84–87, 102, 105–121, 127
Triton X-100	2, 8, 49, 50, 53, 54, 78
Ultraviolet	31, 44
Urine	6, 102
Vitamin C	10
Vorticity	40, 41
Wall effects	53, 54
Westrex teletype	79
X-rays	49–53, 107, 111, 112, 116
Yttrium-88	49, 52
Zinc-65	49, 52